THE ESSENCE OF

ENGINEERING FLUID MECHANICS

Marcel Escudier

Prentice Hall Europe
LONDON   NEW YORK   TORONTO   SYDNEY   TOKYO
SINGAPORE   MADRID   MEXICO CITY   MUNICH   PARIS

First published 1998 by
Prentice Hall Europe
Campus 400, Maylands Avenue
Hemel Hempstead
Hertfordshire, HP2 7EZ
A division of
Simon & Schuster International Group

Typeset in 10 on 12 pt Times
by Mathematical Composition Setters Ltd, Salisbury

Printed and bound by Antony Rowe Ltd, Eastbourne

Library of Congress Cataloging-in-Publication Data

Escudier, M. P.
  The essence of engineering fluid mechanics / Marcel Escudier.
    p.   cm. – (The essence of engineering)
  Includes bibliographical references and index.
  ISBN 0-13-728296-6 (alk. paper)
  1. Fluid mechanics.   I. Title.   II. Series
TA357.E82   1998
620.1'06–dc21
                                                    98-24473
                                                    CIP

British Library Cataloguing in Publication Data

A catalogue record for this book is available from
the British Library

ISBN 0-13-728296-6

THE ESSENCE OF

ENGINEERING FLUID MECHANICS

THE UNIVERSITY OF LIVERPOOL

HAROLD COHEN LIBRARY

For conditions of borrowing, see Library Regulations

THE ESSENCE OF ENGINEERING SERIES

Published titles
The Essence of Solid-State Electronics
The Essence of Electric Power Systems
The Essence of Measurement
The Essence of Engineering Thermodynamics
The Essence of Power Electronics
The Essence of Analog Electronics
The Essence of Optoelectronics
The Essence of Microprocessor Engineering
The Essence of Communications Theory

Contents

Preface

This textbook provides an introduction to fluid mechanics for first-year under-graduates studying mechanical, aeronautical, chemical or civil engineering. Although it is assumed that the majority of readers will have a background in basic physics and mathematics to A-level standard or equivalent, reliance on prior knowledge is slight. The emphasis throughout is on the development of a thorough understanding of a few basic principles which can be applied to calculate such physical quantities of engineering significance as force, pressure, flowspeed, flowrate and power associated with gas and liquid flow through or around a wide range of engineering devices, including engines, turbines, compressors, valves, flowmeters, aircraft, bridges, cars and wings. Since the list of applications is practically endless, the student's aim has to be the understanding of concepts, principles and procedures, not memorising seemingly endless formulae. The few formulae which should be committed to memory are indicated in the text by surrounding each one with a rectangular box. Self-assessment problems have been included at the end of most chapters, almost all based upon recent examination questions set to students at the University of Liverpool. In most cases the problems consist of two parts, the first requiring a derivation of a formula from basic principles, the second a numerical part in which the results of the first are applied to a specific situation. It is the author's belief that the only way to become proficient in fluid mechanics, and indeed any core engineering subject, is by attempting as many such problems as possible with reference to an appropriate text or lecture notes. Rather than learning by rote, it is through this kind of repetitive application that formulae are best committed to memory.

Although there are slight differences from university to university, there is fairly general agreement within the engineering community as to the topics which should be included in a first course of lectures (the word 'module' instead of 'course' is now becoming common) on fluid mechanics. The topics included in this textbook correspond with what has been taught by the author and his colleagues over at least the past decade. Some lecturers may question the length of the chapter on dimensional analysis, its treatment so early in the book and even its inclusion at all (in some universities it is regarded as a second-year topic). The author's view is that dimensional analysis is a topic of such general applicability in engineering and physics that it fully justifies its prominence. Also, thanks to the use of the names of such great engineers and scientists as Euler, Mach and

Reynolds to identify particular non-dimensional groups, dimensional analysis provides a convenient way of providing a historical perspective of fluid mechanics.

Two key topics are treated in a manner which is likely to be unfamiliar to some, possibly most, lecturers. The first is dimensional analysis which we approach using the method of sequential elimination of dimensions. The author believes this technique has clear pedagogical advantages over the widely used Rayleigh exponent method which all too frequently leaves the student with the mistaken idea that all physical processes can be represented by a simple power-law formula. The author also finds the development of the linear momentum equation in Chapter 9 much more straightforward to present to students new to fluid mechanics than the more common approach via Reynolds' transport theorem. The method adopted here shows very clearly the relationship with the familiar $F = ma$ form of Newton's second law of motion and avoids the need to introduce an entirely new concept which is ultimately only a stepping stone to an end result.

There is considerable emphasis on applications to real engineering problems, above all in Chapters 8 and 10 where a wide range of machinery and other equipment is dealt with, from gas and hydraulic turbines to flowmeters and jet pumps.

In keeping with the *Essence of Engineering* series, the scope of this book is intentionally limited, both in the choice of topics and in their theoretical development. Numerous textbooks are already available covering the entire range of topics likely to be treated in the entire time any engineering undergraduate spends at university. Some of these are extremely good and widely used but have evolved into 700-plus page 'blockbusters' which may well overwhelm many students. Much the same can be said of texts which take a far more sophisticated and mathematically elegant approach to the subject. This textbook is intended to serve as an introduction to fluid mechanics for engineering undergraduates generally, but has been written with those students (unfortunately the majority) in mind who find fluid mechanics a difficult subject to master.

Symbols

English symbols		SI unit(s)	Dimension(s)
a	acceleration	m/s^2	L/T^2
A	cross-sectional or surface area	m^2	L^2
B	atmospheric or barometric pressure	bar	M/LT^2
c	concentration	kg/m^3	M/L^3
c	soundspeed	m/s	L/T
Ca	cavitation number		
c_0	speed of light in vacuum	m/s	L/T
C_D	coefficient of discharge		
C_D	drag coefficient		
C_f	friction factor		
C_p	specific heat at constant pressure	m^2/s^2.K	L^2/T$^2\theta$
C_P	pressure coefficient		
C_v	specific heat at constant volume	m^2/s^2.K	L^2/T$^2\theta$
d	diameter	m	L
D	drag force	N	ML/T^2
e	energy	J	ML^2/T^2
E	Young's modulus	Pa	M/LT^2
Eu	Euler number		
F	shear force	N	ML/T^2
F_B	buoyancy force	N	ML/T^2
Fr	Froude number		
g	acceleration due to gravity	m/s^2	L/T^2
G	modulus of rigidity		
h, H	height	m	L
H	horizontal component of force	N	ML/T^2
I	second moment of area	m^4	L^4
I_C	second moment of area about an axis through the centroid	m^4	L^4

j	number of independent dimensions		
k	number of non-dimensional quantities		
K	bulk modulus of elasticity	Pa	M/LT^2
l, L	length	m	L
L	lift force	N	ML/T^2
m	mass	kg	M
$\dot{m}$	mass flowrate	kg/s	M/s
M	Mach number		
M	molar mass	kg/kmol	
M	molecular weight		
M	momentum	kg.m/s	ML/T
$\dot{M}$	momentum flowrate	kg.m/s^2	ML/T^2
n	number of physical quantities		
N	number of molecules		
p	static pressure	Pa	M/LT^2
p_H	hydrostatic pressure	Pa	M/LT^2
p_G	gauge pressure	Pa	M/LT^2
p_0	stagnation pressure	Pa	M/LT^2
p_{REF}	reference pressure	Pa	M/LT^2
p_T	total pressure	Pa	M/LT^2
p_V	vapour pressure	Pa	M/LT^2
$\dot{Q}$	volumetric flowrate	m^3/s	L^3/T
r	number of dimensions		
r, R	radius	m	L
R	reaction force	N	ML/T^2
R	specific gas constant	m^2/s^2.K	$L^2/T^2\theta$
$\mathfrak{R}$	universal gas constant	m^2/s^2.K	$L^2/T^2\theta$
$\mathfrak{R}$	molar gas constant	kJ/kmol.K	$L^2/T^2\theta$
Re	Reynolds number		
s	distance	m	L
S	fluid–structure interaction force	N	ML/T^2
St	Strouhal number		
t	elapsed time	s	T
T	absolute temperature	K	θ
T	temperature	°C	θ
T	surface-tension force	N	ML/T^2
T	thrust	N	ML/T^2
T	torque	N.m	ML^2/T^2
u	velocity	m/s	L/T
v	velocity	m/s	L/T
V	vertical component of force	N	ML/T^2

V	velocity	m/s	L/T
V_D	vertically downwards force	N	ML/T^2
V_U	vertically upwards force	N	ML/T^2
$\bar{V}$	average or bulk mean velocity	m/s	L/T
v	specific volume	m^3/kg	L^3/M
ϑ	volume	m^3	L^3
ϑ_c	critical volume	m^3	L^3
w	specific weight	N/m^3	M/L^2T^2
W	weight	N	ML/T^2
W	width	m	L
We	Weber number		
X	length	m	L
z	depth of liquid	m	L
z'	altitude	m	L
z_C	depth of centroid	m	L
z_P	depth of centre of pressure	m	L
Z	total depth of liquid	m	L

Greek symbols

γ	ratio of specific heats		
δ	gap width	m	L
δA	element of area	m^2	L^2
δF	element of force	N	ML/T^2
δH	element of horizontal force	N	ML/T^2
δm	element of mass	kg	M
δp	infinitesimal change in pressure	Pa	M/LT^2
δs	infinitesimal change in distance	m	L
δt	infinitesimal change in time	s	T
δV	element of vertical force	N	ML/T^2
$\delta \vartheta$	element of volume	m^3	L^3
δW	element of weight	N	ML/T^2
δx	element of length	m	L
δz	element of depth	m	L
δh	height difference	m	L
Δp	pressure difference	Pa	M/LT^2
ΔZ	depth difference	m	L
θ	angle		
κ	lapse rate	°C/m	θ/L
μ	dynamic viscosity	Pa.s	M/LT
ν	kinematic viscosity	m^2/s	L^2/T
Π	non-dimensional number		
ρ	density	kg/m^3	M/L^3

σ	surface tension	N/m	M/T²
σ	relative density		
τ	shear stress	Pa	M/LT²
τ_Y	yield stress	Pa	M/LT²
φ	angle	rad	

Subscripts

C   centroid
F   fluid
G   gas
M   manometer
P   centre of pressure
REF reference
S   surface
s   isentropic
T   total
x   *x*-direction
y   *y*-direction
0   stagnation conditions

CHAPTER 1

Introduction

Why do students of engineering need to study fluid mechanics? First and fore-most, the answer is 'design'. It can be argued that the whole point of engineering is engineering design, and it is the exception rather than the rule where consider-ations of fluid flow are not crucial to the design process. In this chapter we support this statement by indicating the enormous and diverse range of practical situations where fluid mechanics plays a role, often together with related subjects such as heat transfer, thermodynamics and combustion. Although the emphasis of this book is on applications of fluid mechanics in mechanical, aeronautical and civil engineering, we also mention many of the natural phenomena for which fluids and the way they flow play a fundamental role.

By the end of this book the student should be able to

- scale up the results of wind-tunnel model tests using dimensional analysis (Chapter 3): a typical application of dimensional analysis is in the analysis of wind-tunnel data for the aerodynamic behaviour of a Formula One racing car, as shown in Figure 1.1, though we could just as well have chosen a fighter aircraft or a bridge;
- specify the centrifugal pump (illustrated in Figure 1.2) characteristics necessary to handle large quantities of oil based upon small-scale tests with water, again guided by dimensional analysis (Chapter 3);

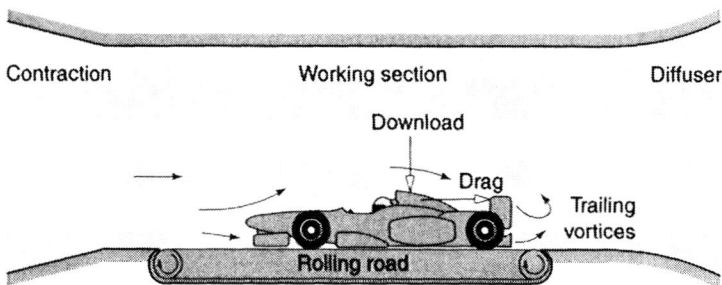

Figure 1.1 *Wind-tunnel test of a Formula One racing car*

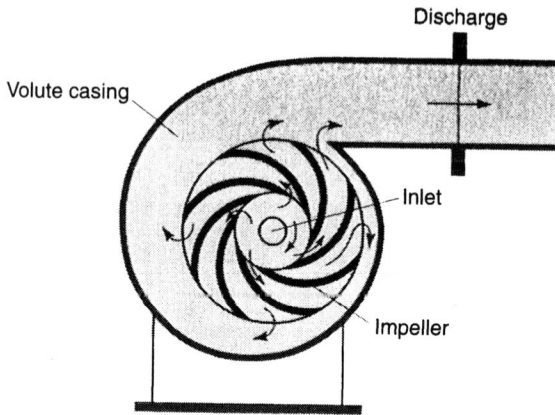

Figure 1.2 *Centrifugal pump testing*

- calculate the flowspeed in a wind tunnel using a Pitot-static tube and a U-tube manometer, as shown in Figure 1.3: this calculation involves both hydrostatics (Chapter 4) and also Bernoulli's equation (Chapters 7 and 8);
- calculate the force exerted by the water of a reservoir on the face of a dam, as shown by R in Figure 1.4, using the principles of hydrostatics (Chapters 4 and 5);
- use the principles of hydrostatics (Chapters 4 and 5) to design a boom to contain an oil slick, as shown in Figure 1.5;

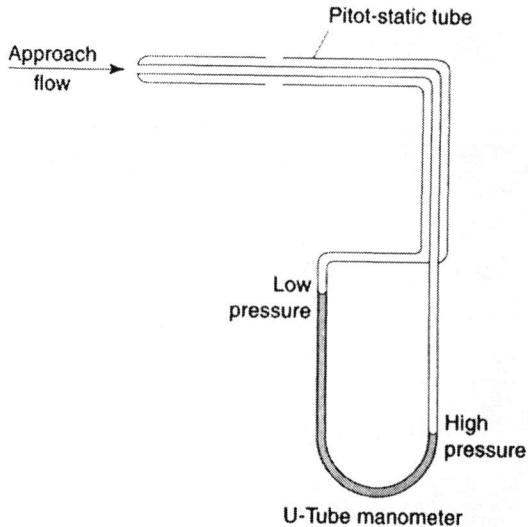

Figure 1.3 *Pitot-static tube and U-tube manometer*

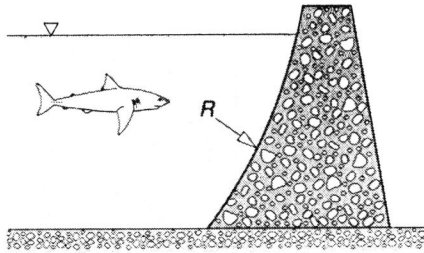

Figure 1.4 *Hydrostatic force on the face of a dam*

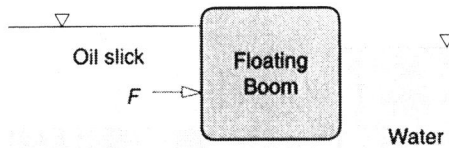

Figure 1.5 *Boom designed to contain an oil slick*

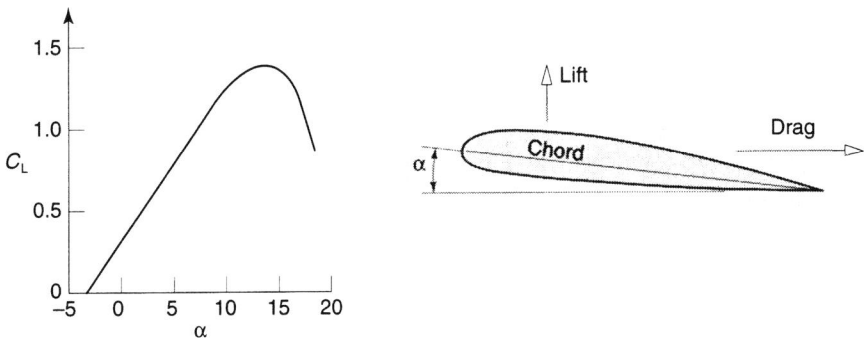

Figure 1.6 *Aerodynamic lift generated by airflow over an aerofoil*

- use Bernoulli's equation (Chapters 7 and 8) to calculate the lift resulting from airflow over the surfaces of an aerofoil, as shown in Figure 1.6;
- use Bernoulli's equation (Chapters 7 and 8) to determine the flowrate at which internal boiling (cavitation) occurs at room temperature in the flow of a liquid through a valve or, as illustrated in Figure 1.7, a convergent–divergent nozzle;
- use the continuity (Chapter 6) and momentum (Chapters 9 and 10) equations to calculate the thrust of a turbofan jet engine, as shown in Figure 1.8;

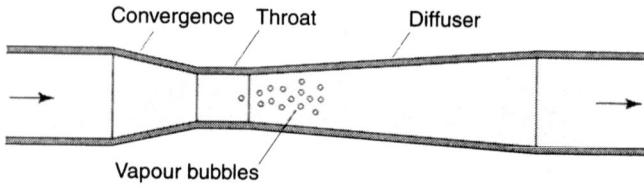

Figure 1.7 *Cavitation in water flow through a nozzle*

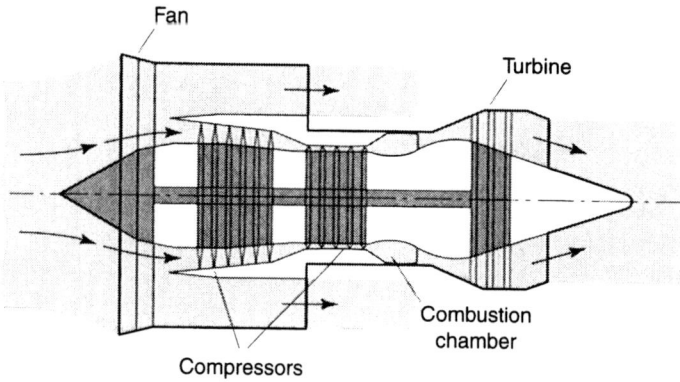

Figure 1.8 *Thrust generated by a turbofan jet engine*

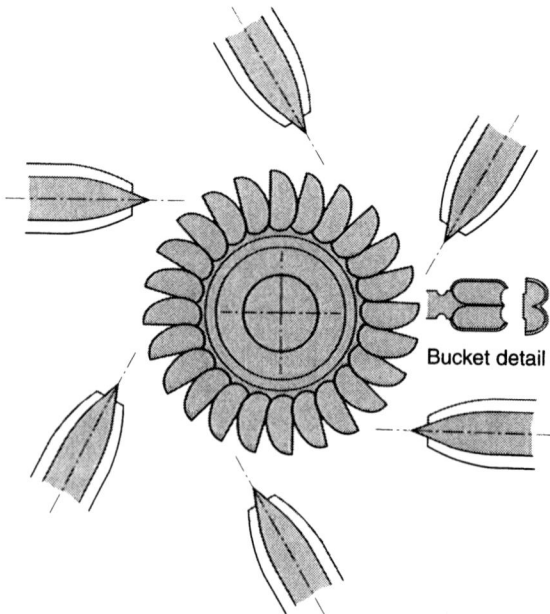

Figure 1.9 *Power output of a Pelton turbine*

- use the continuity (Chapter 6) and momentum (Chapters 9 and 10) equations, together with Bernoulli's equation (Chapters 7 and 8) to calculate the power output of a Pelton hydraulic turbine, as shown in Figure 1.9.

The foregoing is just a selection of the engineering applications considered in this textbook. As we point out in the remainder of this chapter, there are few areas of life, whether man-made or natural, in which fluid mechanics does not play a vital role.

1.1 What are fluids and what is fluid mechanics?

Without water we die within about ten days, and without the oxygen contained in the air we breathe we become brain dead within about four minutes. Water is a liquid, air is a gas, and both are what we call fluids. The total mass of air in the atmosphere which surrounds the earth is about 5.3×10^{18} kg and the total mass of water in all the oceans, rivers, lakes, etc., is about 1.4×10^{21} kg. Given their abundance, and perhaps also their importance to our very existence, it is hardly surprising that these are the two fluids encountered most commonly in fluid mechanics. There are, of course, many other fluids. Methane, ethane, hydrogen, helium, oxygen and nitrogen are all gases which behave much like air. Similarly, oil, petrol, mercury, honey, glycerine and alcohol are all relatively simple liquids much like water, but with different densities and viscosities (terms we explain in Chapter 2 which is concerned with fluid properties and with what makes fluids different from solids). We call these simple fluids *Newtonian*. Blood, synovial fluid (which lubricates our joints), custard, mayonnaise, salad cream, ketchup, hair gel, toothpaste, drilling fluid, freshly mixed cement and paint are also liquids but with viscous properties and flow behaviour very different from those of water. These differences arise primarily because such liquids have either a complex molecular structure or consist of a mixture of a simple liquid and many tiny (often sub-micron size) suspended particles. Because of the complexity of their viscosities, these fluids are termed *non-Newtonian*. The treatment of non-Newtonian liquids is beyond the scope of this textbook: even the study of their viscous properties is a subject in itself, called rheology.

We know from everyday experience that fluids flow. Water flows from the mains supply when we open the tap. Water flows from a sink into the drainage system. Tea flows from a teapot. Beer flows into our bodies from a glass, bottle or can and, usually after a biological transformation, flows out again. Blood flows through our arteries and veins. Air flows into our lungs and carbon dioxide flows out. Petrol and air flow into our car engines and exhaust gas flows out. Town gas, a mixture consisting primarily of hydrogen, methane and carbon monoxide, flows to our gas stoves and boilers and products of combustion flow away. Air flows around us as we walk, run or ride our bicycles. Air flows over the bodywork of our cars and over the wings and fuselages of the aircraft we fly in. Oil, gas and brine flow from deep in the earth to the surface when we drill for oil and gas. Water flows from rivers into reservoirs, lakes and the sea. Water flows around the hull of a ship or a submarine. Lava flows from an active volcano.

The study of fluid mechanics is concerned with calculating the characteristics of fluid flows such as those mentioned in the previous paragraph. As we outlined at the start of this chapter, an understanding of fluid mechanics enables us to answer questions such as:

- What is the fluid velocity and pressure at any point within a flowing fluid?
- What forces does the fluid exert on a surface with which it is in contact?
- How does the shape of the surface affect the flow velocity and pressure?
- How can we measure fluid pressure and flowspeed?
- What power is needed to pump a fluid at a specified flowrate?
- How do we scale up the results of a small-scale experiment to full-scale conditions?

Fluid mechanics is the analysis of fluid flow based upon the law of conservation of mass, Newton's laws of motion and the laws of thermodynamics, together with appropriate representation of fluid properties. However, before we go into further detail, it is useful to expand the catalogue of situations where fluid mechanics plays an essential role.

1.2 Fluid mechanics in nature

The height of the earth's atmosphere is usually taken to be about 80 km and for many purposes the atmosphere can be regarded as a stationary layer of fluid with the temperature variation shown in Figure 1.10. We consider this hydrostatic model of the atmosphere in some detail in Chapter 4. We know, of course, that the atmosphere, especially that part of it we inhabit, is far from static and meteorology is the branch of fluid mechanics devoted to the study of its motion. Anyone who has seen time-lapse film of clouds knows that in addition to being swept along by winds they are in constant motion due to thermals, evaporation, condensation and shearing. Dust devils, tornadoes and hurricanes are examples of the violent swirling fluid motion which can arise in the lower atmosphere due to combined thermal and shearing effects.

While it is essential that the earth is surrounded by a layer of air, it is just as essential for humans (at least in our current state of evolution) that all the water in the hydrosphere is not uniformly distributed over the surface of the earth. Were that the case, the water layer would be about 2.7 km deep. Instead this water actually covers about 71% of the earth's surface with regions of the deepest ocean being about 10 km deep, almost equal in magnitude to the height of Mount Everest. As with the atmosphere, much can be learned about the oceans, reservoirs, lakes, etc., by considering them to be at rest. Chapters 4 and

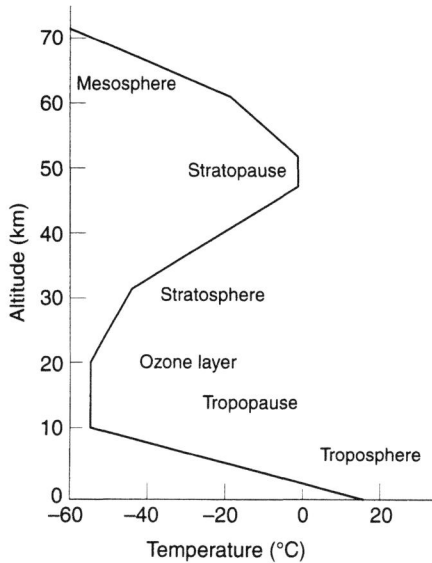

Figure 1.10 *Structure of the atmosphere*

5 are devoted to hydrostatics – the study of fluids at rest – with a considerable fraction concerned with the fluid forces exerted on surfaces such as the face of a dam, as shown in Figure 1.4. Oceanography is the branch of fluid mechanics which deals with tides, currents, waves, stratification (water density variations due to salt and temperature changes with depth) and other phenomena associated with water motion in the oceans. Related topics involving fluid mechanics are erosion, sedimentation, whirlpools and river flows, and also the flow in canals and sewers, although these are man-made rather than natural systems. We could also include here the fluid mechanics associated with the wave-like body motion which fish, eels, aquatic mammals and sperm use to swim.

Undeniably natural are the flows of lava from an active volcano and of hot water and steam from a geyser. The flow of formation fluids (oil, methane, hydrogen sulphide, brine, etc.) from an oilwell, on the other hand, represents a mixture of man-made and natural phenomena. There would be no flow were it not for the man-made well, but the flow of formation fluids through porous rock involves natural fluids flowing through naturally occurring channels in a natural medium.

Biofluidmechanics is concerned with flow in the circulatory, respiratory and other biological systems. As with all natural systems an additional difficulty here is that the flow geometries are neither well defined nor fixed. For example, the arteries and veins are flexible and change in cross-section as the blood pressure increases and decreases with every beat of the heart. To further complicate matters blood is not a homogeneous fluid but consists mainly of tiny red corpuscles, about 8 μm in diameter shaped like discs with a thick rim, suspended in plasma and so has a non-Newtonian viscosity. Blood at rest has a viscosity about one hundred times that of water though this decreases to about five times that of water in the large arteries. In any event, the well-known saying 'blood is thicker than water' is entirely accurate. Although synovia, the fluid which lubricates our joints, is homogeneous its viscosity is at least as complex as that of blood because it has a polymer-like molecular structure which gives it a form of elasticity, and also a viscosity which decreases when the fluid flows.

1.3 **External flows**

As engineers we are most likely to be concerned with fluid which flows either through or around man-made devices, which we term *internal* and *external* flows, respectively. In the case of buildings, smoke stacks, bridges, wind turbines, and windmills the flow is again provided by nature. The damage which occurs to these and other structures when very high windspeeds arise tells us that the wind can impose enormous forces on their surfaces. In certain circumstances, even at relatively low speeds, a steady wind can excite vibrations which can be of sufficient amplitude to cause structural damage. Massive plate-glass windows have been known to pop out of their frames due to wind-induced torsional oscillations of skyscrapers, while the best-known example of wind-induced vibration is the complete destruction in 1940 of the Tacoma Narrows Bridge in Washington State, USA. In order to design structures which are safe, we need to be able to calculate both the steady and periodic forces due to the wind, either from fundamental theory alone or, more likely, from a combination of theory and experiments carried out in a wind or water tunnel. The same considerations have to be given to the design of offshore structures such as drilling platforms. Another aspect of fluid mechanics concerns the dispersion of the plume of pollutants emitted from a smoke stack, and this too would usually involve a combination of theory and model experiments. Dimensional analysis, which is the basis for scaling up model tests to full-size applications, is the subject of Chapter 3.

Some of the most advanced theoretical and experimental work in fluid mechanics has been associated with the development of aircraft, rockets and missiles. There have been quite remarkable advances in aviation since Orville Wright flew a powered heavier-than-air machine 260 metres in 59 seconds in December 1903. For example, we now take for granted the Boeing 747–400 jumbo jet with a passenger-carrying capacity of 448, a take-off weight in excess of 350 000 kg, a wingspan of 64 m, a cruising speed of nearly 1000 km/h and a range of 13 390 km. Just as impressive is the British Aerospace/Aerospatiale Concorde which routinely carries 128 passengers at twice the speed of sound (a flight speed of about 2130 km/h) in the stratosphere. Although, as we see from Figure 1.10, the outside temperature in the stratosphere is −56.5°C (see Chapter 4), the skin of Concorde reaches a temperature of about 120°C causing the length of the aircraft to increase in flight by about 0.3 m! Some combat aircraft, such as the McDonnell Douglas F-15C Eagle, can fly at even higher speeds (Mach numbers greater than 2.5 or about 2700 km/h). Although manned flights into space are now regarded as almost routine, in fact each flight represents an extraordinary engineering achievement. For example, the speed required to escape from the earth's gravitational pull is about 11 km/s, and on re-entry into the earth's atmosphere the air surrounding the space shuttle becomes so hot (6000°C plus) that the craft is surrounded by a glowing plasma.

One of the ways we distinguish between different flight regimes is through the Mach number which is the ratio of the flight speed to the speed of sound (discussed further in Chapters 3 and 6). As the Mach number increases, the fluid mechanics becomes more and more complicated because more and more physical effects have to be taken into account. If the Mach number is considerably less than 1 (0.3 is the value usually taken), changes in fluid density are negligible and the flow is said to be incompressible. For higher Mach numbers compressibility (i.e density changes) become increasingly important but can be accounted for in a relatively straightforward way using the ideal gas law to relate temperature, pressure and density (see Chapter 2) together with the first law of thermodynamics. Once the Mach number exceeds five, however, very high temperatures develop near surfaces and the air properties change due to chemical breakdown of the molecules and subsequent reaction of free atoms. At this point physical chemistry also comes into play.

The preceding paragraph introduces an important aspect of the subject of fluid mechanics which students often find difficult to understand. Even at low flowspeeds there are few problems we can solve completely, usually because the mathematics involved becomes far too complicated even if we know all the physics. To a degree computers can take over at some stage in the analysis of a problem to provide a numerical rather than an analytical (i.e. algebraic) solution. However, even the largest computers presently available are inadequate to solve most practical problems unless we introduce approximations, assumptions and simplifications. In fact this approach represents common sense. For example, if we are dealing with a flow at low speed where we know that the fluid density remains practically constant, there is no point in making our task more difficult and more expensive than necessary by not introducing this simplification from the outset. Of course, it is usually a matter of experience or even hindsight which tells us what simplifications are justified. In this textbook we approach problems using the simplest possible physics and mathematics with the aim of providing some insight into the interplay between fluid velocity, fluid pressure and flow geometry. However, the reader needs to bear in mind that our approach represents only a start rather than a complete treatment of the solution of problems of fluid flow.

Even land vehicles have now reached speeds where density variations must be accounted for. At the time of writing, the jet-powered car Thrust SSC has just achieved a land speed record of 1228 km/h, which corresponds to a Mach number of 1.018. Somewhat slower is the Japanese Bullet train which has a top speed of about 272 km/h or 75.6 m/s, corresponding to a Mach number of 0.22 so that compressibility effects are negligible. However, racing cars are now achieving speeds where compressibility effects cannot be neglected: the highest speed reached at the new oval track in Fontana, California, is about 400 km/h or 111 m/s, which corresponds to a Mach number of 0.33. Although this figure is close to the 0.3 'cutoff', it must be the case that on the wings and bodywork of the cars there will be regions where the airspeed is considerably higher. However, normal cars, buses and trucks have considerably lower top speeds and the airflow around them can safely be considered to have constant density.

Although the flowspeeds for even the fastest marine vehicles are much lower than is the case for most land vehicles, the fluid mechanics is complicated by wave motion which arises due to the tendency for gravitational pull to overcome any disturbance to a water surface. We are all familiar with the surface gravity waves which propagate radially outwards when we throw a stone into a pond, whereas the forward movement of a ship creates a vee-shaped pattern of surface waves. Although invisible to the eye, a submarine travelling deep below the surface also generates gravity waves as it disturbs water layers which occur due to changes in salt content and temperature. The energy required to generate waves is provided by the propulsion system of the ship or submarine and so corresponds to an additional contribution to the drag force.

1.4 Internal flows

Most of the flow situations dealt with in this textbook are concerned with internal flows through pipes, nozzles, engines, turbomachines, etc. In one sense internal flows are easier to deal with than external flows because the boundaries within which flow occurs are fixed unlike the flow over an aerofoil (Figure 1.6), for example, where the region of flow is effectively unlimited.

The most common man-made device through which flow occurs is a metal or plastic pipe of circular cross-section. Pipes of this kind allow oil and gas to flow to the surface from reservoirs thousands of metres below the surface of the earth, often deep below the seabed, and then hundreds of kilometres across land, to refineries or to ports for transfer to ships. Oil and gas pipelines, and also the pipes which convey water into the turbines of hydroelectric power plants, may be a metre or more in diameter. The enormous capital cost means that very careful consideration has to be given to the design of such pipelines and all the associated valves, pumps, monitoring equipment, etc. Smaller diameter pipes connect the pumps, separators, boilers, distillation columns, burners, filters and so on of oil refineries and other chemical processing plants. Such pipes also allow gas and water to be transported to the homes where we live and to the offices and factories where we work. Fluid flow through a pipe is resisted by friction between the fluid and the internal surface of the pipe which arises due to the viscosity of the fluid and has to be overcome by a pressure difference created by a pump or gravitational effects. The friction also causes the fluid temperature to rise, the fluid density to decrease and the fluid velocity to increase. In the case of a gas, the fluid velocity in a pipe may even reach the speed of sound, causing an effect called *choking* which actually limits the quantity of gas which can be pumped through the pipeline. Clearly, then, even a flow which at first sight appears to be one of the simplest we can think of turns out to be rather complicated. In fact the situation is even more complicated than we have indicated so far because it is only at very low flowrates, or for very small pipes, that the flow remains smooth and steady (so-called *laminar* flow) and we are able to analyse it completely. The majority of flows of engineering interest exhibit a high degree of random unsteadiness which we call *turbulence*, and even today we are only on the verge of being able to calculate turbulent pipe flow from first principles through the use of supercomputers. Fortunately the principles of dimensional analysis apply whether a flow is laminar or turbulent and this enables us to generalise experimental data for use in engineering design calculations.

In industrial applications pipes rarely stay straight or keep the same diameter for long. Often more important than understanding the details of the flow within a pipe is the ability to calculate the very considerable hydrodynamic forces which can arise when a pipe changes direction and, perhaps, also changes diameter, as illustrated by the pipe bend in Figure 1.11. The calculation of hydrodynamic forces is one of the main considerations of Chapter 10, the last in this textbook, and this brings together many of the concepts and principles introduced throughout, particularly those in Chapters 6 and 9.

Combustion chambers, furnaces, boilers, jet pumps, control valves, guidevanes, cyclone separators, radiators, oil coolers, carburettors, fuel-injection

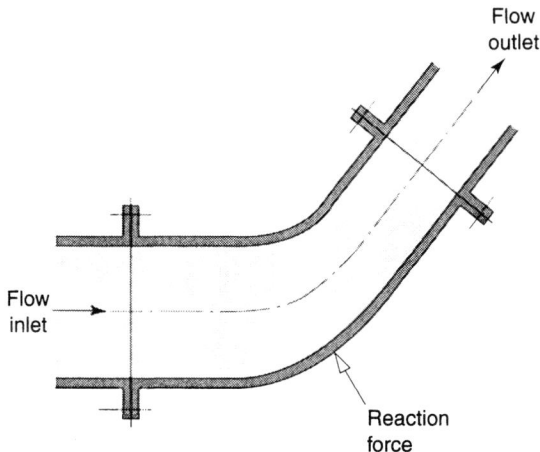

Figure 1.11 *Hydrodynamic reaction force exerted on a pipe bend*

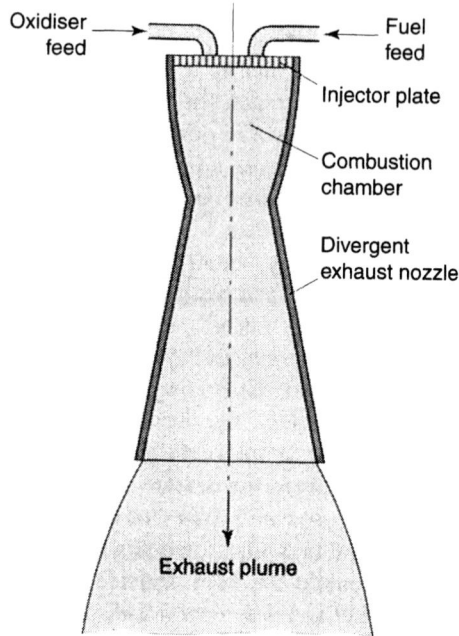

Figure 1.12 *Thrust of a liquid-propellant rocket*

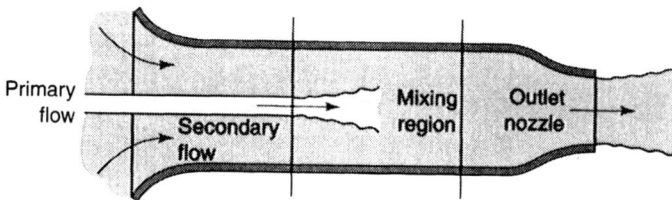

Figure 1.13 *Performance of a jet pump*

systems, rocket engines and the coolant channels within the core of a nuclear reactor, petrol or diesel engine, are all examples involving internal fluid flow in the absence of continually moving parts. As we show in Chapter 10, the flow characteristics which are basic to the design of many of these devices, including the rocket engine, jet pump and cascade of guidevanes, shown in Figures 1.12, 1.13 and 1.14, can be determined using the principles of fluid mechanics that we cover in this textbook. The analysis of most of the other cases requires more advanced aspects of fluid mechanics and may also involve consideration of heat transfer, thermodynamics and chemistry.

The turbojet and turbofan jet engines shown in Figures 10.3(a) and 1.8 are

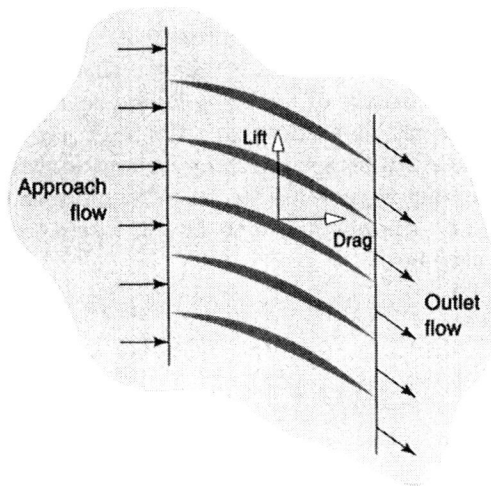

Figure 1.14 *Hydrodynamic forces on a cascade of guidevanes*

examples of a class of devices called *turbomachines*, derived from the Latin word *turbo* which has the meanings 'whirlwind' and 'spinning top'. Other turbomachines are pumps, fans, compressors, gas turbines, steam turbines, hydraulic turbines (see Figure 1.9), turbochargers and superchargers. A common feature of all turbomachines is a central rotating shaft which carries blades (the *rotor* or *impeller*) to transfer momentum and work either to or from the fluid which flows through the machine by causing changes in the direction of fluid flow. Most turbomachines also incorporate stationary blades (called *stators* or *nozzle rings*) attached to the casing to guide the flow to and from the stator stages. As we show in Chapter 10, we can learn a considerable amount about the performance of these very complex machines simply by considering the state of the flow at inlet and outlet. The basic flow within a stator or rotor stage can be analysed in much the same way as that through the cascade of guidevanes, but to take the analysis further once again requires that we use more advanced principles of fluid mechanics, often together with considerations of thermodynamics.

1.5 Summary

In this chapter we have indicated the wide array of engineering devices from the kitchen tap to supersonic aircraft, the basic design of which depends upon considerations of the flow of gases and liquids. Much the same is true of most natural phenomena from our weather to ocean waves and the movement of sperm. This textbook introduces a number of the concepts, principles and procedures which underly the analysis of any problem involving fluid flow. In this Introduction, we have selected fourteen examples for which the student should be in a position to make useful engineering design calculations by the end of the book. This outcome represents a major return on investment for the student, given that the material of the book can be covered in about eighteen hours of lectures (twenty-four 45-minute lectures)! As a final comment, however, we emphasise that simply attending lectures (or reading this book) is not sufficient: it is absolutely essential for the student to spend at least twice the amount of lecture time attempting to solve the self-assessment problems which follow most of the chapters.

CHAPTER 2

Fluids and fluid properties

Wet. Sticky. Viscous. Viscid. Gelatinous. Slippery. Greasy. Oily. Lubricious. Slimy. Oleaginous. Oozy. Soapy. Thick. Thin. Runny. Syrupy. Treacly. Tacky. Claggy. Muddy. Gummy. Gooey. Mucilaginous. Glutinous. Each of these adjectives, commonly used to describe liquids, convey something about how liquids feel, how they flow or how they respond to being stirred or mixed. The list of words used to describe gases is far more limited: viscous, viscid, heavy and dense. We could also include smelly in both lists. In this chapter we introduce the properties used to quantify the physical characteristics of liquids and gases, in contrast to these adjectives which give us primarily a qualitative tactile impression. We discuss how and why fluids and solids are different both on a molecular and on a macroscopic scale. We show that central to the definition of the physical properties of fluids, and the way in which we go on to analyse fluid flow, is the continuum hypothesis which allows us to define properties on a scale which is far smaller than any scale of engineering interest but still far larger than the underlying molecular scale.

2.1 **Fluids and solids**

The state of any substance can be classified as *solid* or *fluid*, with the term fluid including both liquids and gases. From an engineering point of view, the essential difference between a fluid and a solid is the way in which the substance resists shear stress. In the case of a solid, the shear stress is resisted by a static deformation, the magnitude of which (for a given shear stress) depends upon a material property called the *modulus of rigidity*. For a fluid, no matter how low the shear stress, the deformation increases without limit as long as the shear stress is applied. The rate of deformation changes according to a fluid property called the *viscosity*.

We can begin to quantify the statements in the preceding paragraph as follows. Suppose we have a solid rectangular block subjected to a shear (i.e. tangential) force, as illustrated in Figure 2.1(a). Unless the magnitude of the force is so great that the material breaks, the solid resists the force F by a static deformation which we can measure by the angle ϕ (phi). In the case of an elastic solid, according to Hooke's law, the deformation is proportional to the applied shear force so that we can write

$$\frac{F}{A} = \tau = G\phi \qquad (2.1)$$

where A is the surface area over which F is distributed, τ (tau) is the shear stress (i.e. shear force per unit area) and the constant of proportionality G is called the modulus of rigidity or shear modulus.

Consider now the situation illustrated in Figure 2.1(b) which shows a fluid between two parallel plates separated by a distance h, with the lower plate stationary and the upper plate moving at speed V. A fundamental concept of fluid mechanics, called the *no-slip condition* (see section 6.4) is that fluid in contact with a surface sticks to it and moves at the speed of the surface. Thus the fluid in the immediate vicinity of the upper surface moves forward at speed V, the fluid in contact with the lower surface is at rest, and the fluid in between moves as though in infinitesimally thin layers with speed u, which increases progressively with distance y from the lower surface, i.e.

$$u = \frac{Vy}{h}. \qquad (2.2)$$

If we imagine a vertical line marking the fluid at some instant of time, a time t later the line would have rotated through an angle ϕ, as shown in Figure 2.1(b), so that

$$\tan \phi = \frac{Vt}{h}.$$

If the time t is very short, the angle ϕ will be small and negligibly different (measured in radians) from $\tan \phi$, so that

$$\phi = \frac{Vt}{h}$$

from which we see that if t doubles, ϕ also doubles. If t triples, ϕ also triples. And so on. Rather than consider deformation continually increasing in this way, it is far more convenient to think in terms of the rate of change of the deformation, which is given by

$$\frac{d\phi}{dt} = \frac{V}{h}.$$

From equation (2.2) we can see that the quantity V/h is the gradient of the velocity u with respect to the distance y, so that

$$\frac{du}{dy} = \frac{V}{h}.$$

Because gradients of velocity in a fluid occur due to the effect of shear stresses, the rate of deformation du/dy is usually referred to as the *shear rate*. The equivalent statement to equation (2.1) for a fluid can now be written:

$$\tau = \mu \frac{d\phi}{dt} = \mu \frac{du}{dy} \tag{2.3}$$

where the symbol μ (mu) represents the fluid property known as *viscosity* (in some books the symbol η (eta) is used rather than μ). Viscosity is the principal property which distinguishes a fluid from a solid, and many of the adjectives listed at the start of this chapter are qualitative descriptions of the viscous nature of fluids. For many simple fluids, including air and water, μ is a thermodynamic property which depends only upon temperature and pressure but not on the shear rate; such fluids are known as Newtonian fluids after the great English scientist Isaac Newton (1642–1727). One of Newton's many contributions to scientific understanding was the recognition that the resistance to relative motion between two parts of a fluid is proportional to the velocity difference between the parts.

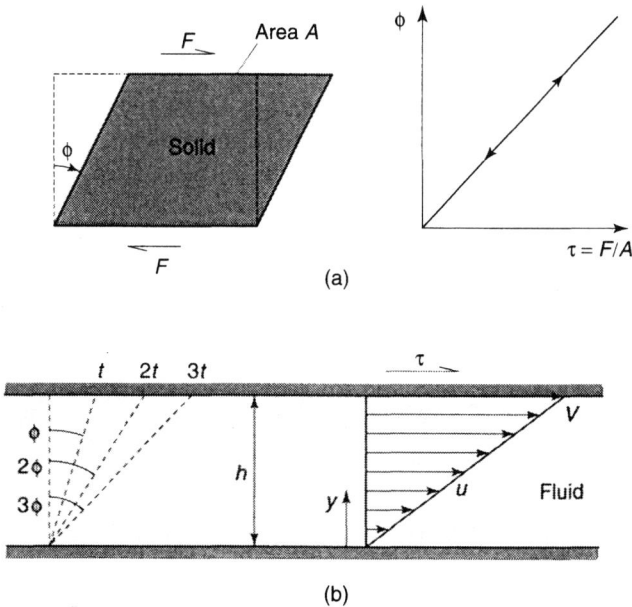

Figure 2.1 *Shear force applied to (a) elastic solid (b) fluid*

It is easy to find descriptive distinctions between the three states in which matter occurs. Solids are hard and not easily deformed. A liquid has no inherent shape and is so easily deformed that due to the influence of gravity it simply takes up the shape of any container into which it is poured without a change in volume. A gas is even easier to deform than a liquid and increases in volume without limit unless constrained by a closed container which it then fills completely. The volume of a fixed mass of gas is decreased by any increase in pressure, whereas to decrease the volume of a liquid by a measurable amount requires very high pressures. These and other differences between the gas, liquid and solid states can be explained on the basis of their molecular structures. Movement of the molecules of a solid is highly restricted because they are closely packed in a specific lattice formation with large intermolecular cohesive forces between them. The molecules of a liquid have more freedom of movement, but because they are further apart (though the typical spacing is still only 10^{-10} m, or 0.1 nm) and the intermolecular forces are therefore smaller, they are in a continual state of interaction with their neighbours and so never move very far. Gas molecules, on the other hand, move about in straight lines at high speed (about 1.2 to 1.5 times the speed of sound, which we discuss in section 2.11 below), occasionally colliding with each other or the surface of a confining container. For both liquids and gases, the *continual bombardment of any surface by molecules gives rise to a stress normal to the surface which we call pressure.*

Since many substances can exist in any one of the three basic states, the differences in molecular structure are largely a matter of degree and there is the possibility of transition between states. For example, the volume of a fixed mass of gas is easily decreased by increasing its pressure, a process called *compression*. At very high levels of compression the gas molecules are forced so close together that the gas becomes indistinguishable from a liquid and is said to liquefy. In fact, the free surface of any liquid is always in contact with its gaseous state, called a *vapour*. It is not only pressure which influences the state of a substance. At a sufficiently high temperature many solids melt and become liquid and, with a further increase in temperature, increasing amounts of vapour are produced until all the material is in the gaseous state. These different states are identified thermodynamically as phases which represent forms of matter which are physically and chemically stable.

2.2 **Fluid density**

The *density* ρ (rho) of a fluid, sometimes referred to as its *mass density*, is the ratio of the mass m of a given volume of fluid to the volume ϑ,* i.e.

$$\rho = \frac{m}{\vartheta}.$$

(2.4)

* Boxes have been placed around the few equations the reader should commit to memory.

In the SI system of units, which we use exclusively in this textbook and present in some detail in Chapter 3, the units of mass are the kilogram (kg), those of volume the metre^3 (m^3) and so the units of density are kg/m^3. As we indicated in section 2.1, we can decrease the volume of a fixed mass of gas by increasing its pressure. According to equation (2.4) the consequence of compression is an increase in the gas density. The pressure of the air flowing through a jet engine, such as that illustrated in Figure 1.2, is increased substantially as it passes through the compressor stages and so the air density also increases. It should be evident by now that our definition of density at the start of this paragraph is incomplete: the idea that the density of air can vary with location in a flow implies that (like all fluid properties) we regard density as having a value at any given point. A more complete definition of density requires that the volume ϑ, and hence the mass m, is so small that there is no appreciable variation of density ρ within the volume. At the extreme, we could define a volume so small that at any instant of time it contained a single molecule. Obviously this does not lead to a sensible definition of density but, by progressively increasing the volume above such a low value, we eventually reach a situation where, although molecules will be continually moving into and out of the volume at its boundary, the number of molecules within the volume at any instant will be constant. The effect on the ratio $m : \vartheta$ of progressively increasing ϑ is shown schematically in Figure 2.2. The horizontal scale is highly compressed to the right of the broken vertical line, and highly expanded to its left. Once ϑ exceeds the critical value ϑ_C, we can define a density as a thermodynamic property (i.e. a property which depends only upon temperature and pressure) which is independent of volume and which may vary smoothly and continuously throughout the entire fluid volume. The densities of water, air and other fluids of engineering interest are given in Tables A.2 to A.5.*

There are two principal ways in which the density of a fluid influences flow.

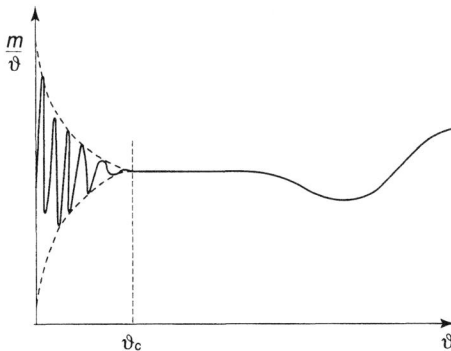

Figure 2.2 *Variation of the ratio mass : volume with volume*

* All tables are to be found in an appendix at the end of the book.

The most important stems from Newton's second law of motion which tells us that the acceleration of a given mass is proportional to the net force applied to it. We shall discuss in detail the application of Newton's second law to fluid flow in Chapters 7 to 10. For the time being it is sufficient to realise that to produce a change in the velocity of a high-density fluid involves much larger forces (per unit volume) than is the case for a fluid of low density. For example, the power required to propel a submerged submarine would be about a thousand times greater than for an airship of the same size and speed flying through the air. The second way in which density plays a key role involves gravity and the associated decrease in atmospheric pressure with altitude or increase in pressure with liquid depth. These and other hydrostatic effects are the subject of Chapters 4 and 5.

2.3 Continuum hypothesis

Since the size of the critical volume ϑ_C is determined by molecular behaviour, it will be different for every fluid and also change with pressure and temperature. However, as we shall now show, ϑ_C will always be far smaller than any volume of engineering interest, except in extreme circumstances, and can be regarded as defining what we mean by a point.

We calculate the average number of molecules contained in a cube of fluid of side length L m. If the fluid density is ρ, from equation (2.4) the mass of the cube will be ρL^3 since the cube volume $\vartheta = L^3$. For a substance of molecular weight M, the mass of 1 kmol of that substance is M kg, so that our cube contains $\rho L^3/M$ kmol (the unit kmol is often written as kg mole). We know that the number of molecules in 1 kmol of any substance is 6.022×10^{26} (Avogadro's number) so that the average number of molecules N in the cube must be given by

$$N = 6.022 \times 10^{26} \frac{\rho L^3}{M} \tag{2.5}$$

which we can use to calculate the average number of molecules in cubes of different sizes for a range of fluids. We start with air for which the molecular weight is 28.97 and, at 0°C and 1 atm (Standard Temperature and Pressure, or STP), the density is 1.294 kg/m^3. From equation (2.5) we obtain the following results:

L	N
1 mm	2.7×10^{16}
1 μm	2.7×10^{7}
100 nm	2.7×10^{4}
50 nm	3362
33.4 nm	1000
20 nm	215
10 nm	27
3.34 nm	1

We cannot give a precise value, but would probably not want the number of molecules over which to form an average to be any lower than 1000 and so conclude that, for air at STP, the concept of fluid density will begin to fail if the cube size is below about 30 nm (i.e. 0.3 μm or 3×10^{-7} m). To put this in perspective, the diameter of a human hair is about 100 μm (i.e. 0.1 mm) and the wavelength of visible light is about 589 nm (0.589 μm). There are few, if any, practical situations involving flow channels with dimensions which come anywhere close to 30 nm, so that assuming we can define fluid density at a point is certainly justified. Gases for which the number of molecules in a 1 μm cube falls below about 1000 are said to be rarefied. Although we used air to illustrate the point here, in fact the results are valid for all gases because for a given temperature and pressure the density of any gas is proportional to its molecular weight.

EXAMPLE 2.1
Hydrogen, which has a molecular weight of 2.02, has a density of 0.090 kg/m^3 at Standard Temperature and Pressure. Calculate the size of a cube of hydrogen which contains an average of 1000 molecules.

SOLUTION
$M = 2.02$; $\rho = 0.090$ kg/m^3; $N = 1000$

From equation (2.5) we have

$$L^3 = \frac{MN}{6.022 \times 10^{26}\rho}$$

$$= \frac{2.02 \times 10^3}{6.022 \times 10^{26} \times 0.090}$$

$$= 3.72 \times 10^{-23} \text{ m}^3$$

from which

$$L = 3.34 \times 10^{-8} \text{ m} \quad \text{or} \quad 33.4 \text{ nm,}$$

the same result as for air.

As we shall see in section 4.12, the air density in the atmosphere decreases with altitude. At the outer limit of the stratosphere (an altitude of about 20.1 km), for example, the density has fallen to 0.088 kg/m^3 and at 70 km, the outer limit of the mesosphere, the density is only 8.28×10^{-5} kg/m^3. The sizes of cubes containing 1000 air molecules corresponding to these densities are 82 nm and 0.83 μm, respectively, which are still extremely small. As we can show from equation (2.5), the density of air would have to fall to 4.8×10^{-14} kg/m^3 for the cube size to reach 1 mm. Only in outer space would such a low density be encountered in nature: the effective air density at an altitude of 1600 km is about 10^{-15} kg/m^3.

Because the molecular structure of liquids is more complex than that of gases, the *number of molecules per unit volume, which is termed the molecular number density,* varies somewhat from liquid to liquid. However, from the values of N calculated from equation (2.5) for cubes of side 1 μm, we see that the molecular number density for liquids far exceeds that for gases:

Liquid	*Number of molecules in a 1 μm cube (volume 10^{-18} m^3)*
Petrol	4.4×10^9
Carbon tetrachloride	6.3×10^9
Liquid oxygen	7.7×10^9
Glycerol	8.2×10^9
Ethyl alcohol	1.0×10^{10}
Water	3.35×10^{10}
Mercury	4.07×10^{10}

Similar considerations to the foregoing allow us to define point values of any fluid property in terms of physical quantities which are measurable on a scale far greater than the molecular scale and which vary smoothly and continuously throughout the fluid. Although these large-scale (macroscopic) properties of a fluid reflect the underlying molecular structure, it is generally the case that we can treat problems of fluid flow without needing to consider molecular structure directly. The idea that both fluid properties and flow properties can be treated in this way is known as the *continuum hypothesis.*

2.4 Specific volume, relative density and specific weight

In thermodynamics it is more usual to work in terms of *specific volume v* than density ρ. The word 'specific' here means 'per unit mass', i.e.

$$v = \frac{\vartheta}{m} = \frac{1}{\rho},$$

from which we see that the units of v are m^3/kg.

Relative density σ (sigma) is the ratio of the density of a substance to that of a standard reference fluid ρ_{REF}:

$$\sigma \equiv \frac{\rho}{\rho_{REF}}.$$

Because it is defined as the ratio of two physical quantities with the same units, relative density has a purely numerical value without units (see Chapter 3). The term *specific gravity* is sometimes used instead of relative density.

For liquids, the reference fluid is usually water at 4°C and 1 atm which has a density of 1000 kg/m³, so that

$$\sigma \ (\text{liquid}) = \frac{\rho}{1000}.$$

The temperature for the reference fluid is sometimes taken as 20°C at which the density of water is 998.20 kg/m³.

For gases, the reference fluid is usually air (though sometimes hydrogen is used) which has a density of 1.204 kg/m³ at 20°C and 1 atm, so that

$$\sigma \ (\text{gas}) = \frac{\rho}{1.204}.$$

In practice, relative density is little used for gases.

Specific weight w, which should not be confused with specific gravity, is weight per unit volume. Since density is mass per unit volume, it follows that

$$w = \rho g$$

where g is the acceleration due to gravity and has the value 9.807 m/s², usually rounded to three significant figures as 9.81 m/s². The units of w can be shown to be N/m³ because, as we shall see in Chapter 3, 1 newton (N) = 1 kg.m/s².

EXAMPLE 2.2

Calculate the density and specific weight for liquid oxygen which has a relative density of 1.46 at −252.7°C, 1 atm.

SOLUTION

$$\sigma = 1.46$$
$$\rho = 1000\sigma$$
$$= 1000 \times 1.46 = 1460 \ \text{kg/m}^3.$$
$$w = \rho g$$
$$= 1460 \times 9.807 = 14 \ 320 \ \text{N/m}^3.$$

Comment: In any problem where either relative density or specific weight are specified, the first step should always be to calculate the fluid density in SI units.

2.5 Ideal-gas law

At very high temperatures (above about 1000°C) the molecular structure of a gas breaks down (a process known as dissociation) and at very high pressures, as we have already indicated in section 2.1, gases can liquefy. Away from these extremes, all gases are in good agreement with the ideal-gas law (also referred to as the perfect-gas law)

$$p = \rho RT \qquad (2.6)$$

where p is the gas pressure in pascal ($Pa = N/m^2$), T is the absolute temperature in kelvin ($K = 273.15 + °C$) and R is a constant of the gas called the specific gas constant (units $m^2/s^2.K$). Equation (2.6) is called a *thermal equation of state*. The specific gas constant is related to the universal (molar) gas constant $\Re$ as follows:

$$\Re = MR = 8314.510 \ m^2/s^2.K \qquad (2.7)$$

where M is the molecular weight. The units $m^2/s^2.K$ suggest a connection between R and a speed, which we shall show in section 2.10 is that for the propagation of sound through the gas, i.e. the speed of sound.*
   The specific gas constant is related to the difference in the specific heats at constant pressure C_p and constant volume C_v, i.e.

$$R = C_p - C_v.$$

Values for the molecular weight M, the specific gas constant R and the ratio of the specific heats

$$\gamma = \frac{C_p}{C_v}$$

are tabulated in Table A.5 for a range of gases. Although values for the corresponding gas density ρ at STP (20°C, 1 atm) are also tabulated, this is not essential since the density of an ideal gas at any temperature and pressure can be calculated from equations (2.6) and (2.7).

---

* Directly equivalent units to $m^2/s^2.K$ for R are J/kg.K, where the symbol J stands for the unit for energy, the joule. However, the units for R are also often stated to be J/kmol.K. For consistency M must then be given the units kg/kmol and referred to as the molar mass. The reader should also be aware that the value for $\Re$ can also be given as 8.314 510 kJ/kg.K, the factor of 1000 being absorbed into kJ.

EXAMPLE 2.3
Calculate the density of nitric oxide (NO) at 20°C and 1 atm and also at
500°C and 5 bar.

SOLUTION
$M = 30.01$; $p_1 = 1.01325 \times 10^5$ Pa; $T_1 = 293.15$ K;
$p_2 = 5 \times 10^5$ Pa; $T_2 = 773.15$ K

From equation (2.7)

$$R = \frac{\Re}{M} = \frac{8314.51}{30.01} = 277.1 \text{ m}^2/\text{s}^2.\text{K}$$

From equation (2.6)

$$\rho_1 = \frac{p_1}{RT_1} = \frac{1.01325 \times 10^5}{277.1 \times 293.15} = 1.248 \text{ kg/m}^3$$

$$\rho_2 = \frac{p_2}{RT_2} = \frac{5 \times 10^5}{277.1 \times 773.15} = 2.334 \text{ kg/m}^3.$$

Comment: It is generally unnecessary to carry so many significant figures
(s.f.) in an engineering calculation. 4 s.f. for $\Re$ and 3 s.f. for p and T are
usually sufficient.

The values calculated here for R and ρ_1 are exactly the same as those given for
NO in Table A.5.
   Here, as in all examples, the first step in the solution was to restate the data
given in standard SI units. The student should develop the habit of converting
given data to standard SI form in this way.

2.6 **Equation of state for liquids**

Although equation (2.6) has no generally valid equivalent that is applicable to
liquids, it is usually adequate to assume that ρ = constant and $C_p = C_v$ = constant,
so that

$$\gamma = \frac{C_p}{C_v} = 1.$$

The effect of extreme pressure on the density of a liquid is given by the approximate expression

$$\frac{p}{B} = (1 + C)\left(\frac{\rho}{\rho_B}\right)^n - C$$

where

$$B = 1 \text{ bar}$$
$$\rho_B = 1000 \text{ kg/m}^3$$

p is measured in bar and C and n are constants for a given liquid.
    For water

$$C = 3000 \qquad \text{and} \qquad n = 7.$$

For further information about equations of state and thermodynamic properties, the reader should consult textbooks on thermodynamics.

2.7 Dynamic viscosity

In section 2.1 we introduced viscosity as the property which provides the link between the shear stress applied to a fluid and the resulting rate of deformation, which we showed was equal to the velocity gradient in the case of a fluid between two plates, one fixed and the other moving. In most flows the velocity variation within the fluid is more complicated than the linear variation shown in Figure 2.1. In more general situations, such as that shown in Figure 2.3, the continuum hypothesis allows us to relate the shear stress τ at any point in the fluid to the velocity gradient at that point according to

$$\tau = \mu \frac{du}{dy}.$$

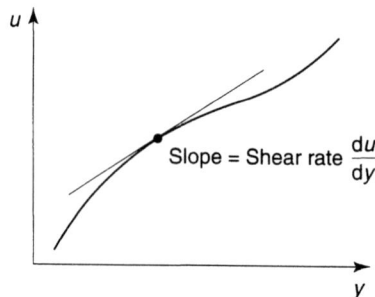

Figure 2.3 *Velocity gradient*

The quantity μ has the units Pa.s ($= N.s/m^2$) and is more properly known either as the absolute coefficient of viscosity or the dynamic viscosity, but is usually referred to simply as the viscosity. The inverse of μ (i.e. $1/\mu$) is called the *fluidity*.

The viscosity of Newtonian liquids is commonly measured using an instrument such as the concentric-cylinder viscometer illustrated schematically in Figure 2.4. The liquid is introduced into the annular gap between an inner cylinder, which rotates at a speed N rpm, and an outer stationary cylinder. The fluid velocity at the surface of the outer cylinder is then zero while that at the surface of the inner cylinder is given by

$$V = \frac{2\pi NR}{60},$$

the factor of $2\pi/60$ being introduced to convert from rotational speed measured in rpm to angular velocity in rad/s. By making the width δ of the annular gap between the two cylinders negligibly small in comparison with the cylinder radius R, the effect of curvature becomes unimportant and the flow geometry is essentially the same as that between two flat surfaces. The velocity gradient

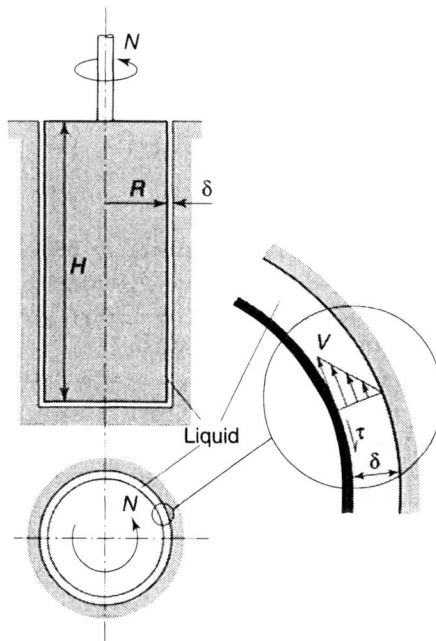

Figure 2.4 *Concentric-cylinder viscometer*

within the fluid is then

$$\frac{du}{dy} = \frac{V}{\delta}$$

and the shear stress

$$\tau = \mu \frac{du}{dy} = \frac{\mu V}{\delta}.$$

The total torque T exerted on the inner cylinder is given by

$$T = \tau A R$$

where A is the surface area of the inner cylinder. If the length of the cylinder is H, we have

$$A = 2\pi R H$$

and so

$$T = \frac{4\pi^2 N R^3 H \mu}{60\delta}.$$

Since δ, R and H are known, μ can be determined by measuring the torque T and the rotational speed N, i.e.

$$\mu = \frac{15\delta T}{\pi^2 N R^3 H}.$$

EXAMPLE 2.4
A concentric-cylinder viscometer is used to measure the viscosity of a viscous oil. The dimensions of the viscometer are $R = 20$ mm, $H = 75$ mm, $\delta = 0.1$ mm. At a rotational speed of 3 rpm the torque is 0.01 N.m. Calculate the dynamic viscosity of the oil.

SOLUTION
$R = 2 \times 10^{-2}$ m; $H = 7.5 \times 10^{-2}$ m; $\delta = 1 \times 10^{-4}$ m;
$N = 3$ rpm; $T = 0.01$ N.m

$$\mu = \frac{15\delta T}{\pi^2 N R^3 H}$$

$$= \frac{15 \times 10^{-4} \times 0.01}{\pi^2 \times 3 \times 2^3 \times 10^{-6} \times 7.5 \times 10^{-2}}$$

$$= 0.844 \text{ Pa.s}$$

Values for the viscosities of a wide range of Newtonian fluids are given in Tables A.4 (liquids) and A.5 (gases) at Standard Temperature and Pressure. The strong dependence of viscosity on temperature is evident from Figure 2.5 whereas the dependence on pressure is generally negligible up to 10 bar for gases and 100 bar for liquids. For gases, the temperature dependence is well represented by Sutherland's formula:

$$\mu = \frac{KT^{3/2}}{T + C}$$

where K and C are constants characteristic of the particular gas. As the following example illustrates, only one constant is needed if (as is usually the case) μ is known at a specified temperature.

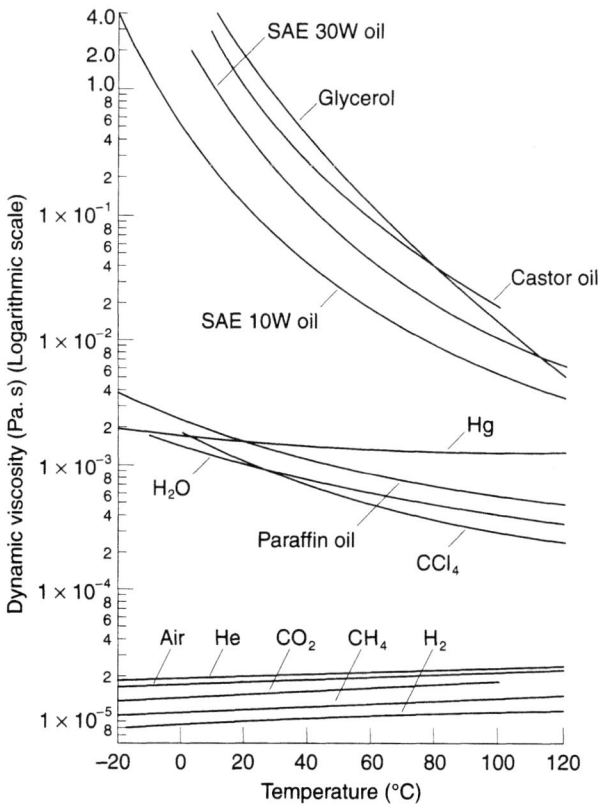

Figure 2.5 *Dynamic viscosity of common fluids as a function of temperature*

EXAMPLE 2.5

For air, $C = 110.4$ K and the viscosity at 20°C is 1.8×10^{-5} Pa.s. Calculate the value of μ at 400°C using Sutherland's formula.

SOLUTION

$T_1 = 20 + 273 = 293$ K; $\mu_1 = 1.8 \times 10^{-5}$ Pa.s; $C = 110.4$ K;
$T_2 = 400 + 273 = 673$ K

From Sutherland's formula

$$\mu_1 = \frac{K T_1^{3/2}}{T_1 + 110.4}$$

so that

$$K = \frac{\mu_1(T_1 + 110.4)}{T_1^{3/2}}$$

$$= \frac{1.8 \times 10^{-5} \times 403.4}{293^{3/2}}$$

$$= 1.448 \times 10^{-6} \text{ Pa.s/K}^{1/2}.$$

At 400°C, Sutherland's formula gives

$$\mu_2 = \frac{1.448 \times 10^{-6} \times 673^{3/2}}{783.4}$$

$$= 3.23 \times 10^{-5} \text{ Pa.s}.$$

Comment: The value calculated for μ at 400°C is within 1% of the standard value given in Table A.3.

For liquids the temperature dependence of viscosity can be approximated by the formula

$$\mu = \mu_0 \exp\left[C\left(\frac{T_0}{T} - 1\right)\right]$$

where μ_0 is the viscosity at temperature T_0 and C is a constant for the particular liquid. The presence of the exponential indicates a strong temperature dependence: for water, for example, the viscosity just above the freezing point 0°C is 1.787×10^{-3} Pa.s compared with 2.818×10^{-4} Pa.s just below the boiling point 100°C, which corresponds to a decrease of 84% or a ratio of 6.3 : 1.

EXAMPLE 2.6
For a typical engine oil, $C = 17$ and the viscosity at 20°C is 1.0 Pa.s. Calculate the viscosity at 150°C.

SOLUTION
$C = 17;$ $T_0 = 20 + 273 = 293$ K; $\mu_0 = 1.0$ Pa.s; $T = 150 + 273 = 423$ K
At $T = 150$°C

$$\mu = 1.0 \exp\left[17 \times \left(\frac{293}{423} - 1\right)\right]$$

$$= 5.38 \times 10^{-3} \text{ Pa.s.}$$

2.8 Kinematic viscosity

The kinematic viscosity ν (nu), defined by

$$\nu \equiv \frac{\mu}{\rho},$$

is frequently used instead of the dynamic viscosity μ because, in many problems, μ and the density ρ only occur in the combination μ/ρ. An interesting consequence of combining μ and ρ in this way is that the kinematic viscosity of many gases is higher than that of many liquids, a trend which becomes more pronounced as the temperature increases because the dynamic viscosities of liquids increase whereas those for gases decrease. The term *kinematic* is associated with the units of ν, which are m^2/s, and so involve only metres and seconds like displacement (m), velocity (m/s) and acceleration (m/s^2) which are descriptive terms not directly involving the dynamics (i.e. the stresses and forces) of a problem. Chapter 6 is concerned with the kinematics of fluid flow.

2.9 Non-Newtonian liquids

Whether a fluid is Newtonian or non-Newtonian, the relationship between shear stress and shear rate is the same, i.e.

$$\tau = \mu \times \text{shear rate.}$$

In section 2.1 we identified Newtonian fluids as those for which the viscosity μ can be regarded as a thermodynamic property which may depend upon temperature and pressure but is independent of any deformation of the fluid as it flows, i.e. independent of the shear rate. Gases have a simpler molecular structure than liquids and always exhibit Newtonian characteristics. However, as we indicated in Chapter 1, the molecular structures of most synthetic liquids, as well as such naturally occurring liquids as blood and synovia, are complex and in consequence the viscosities (sometimes the term *apparent viscosities* is used) of these fluids change not only with temperature and pressure but also with the shear rate itself. Such fluids are termed non-Newtonian. As we shall illustrate, μ can both increase and decrease with shear rate.

Blood, cement slurry, yoghurt, toothpaste, mud, salad cream and many other non-Newtonian liquids of practical importance are shear thinning, which means that their viscosities decrease with increase in the shear rate. The term *pseudoplastic* is also used for shear-thinning fluids. Salad cream and tomato ketchup are often reluctant to leave the bottle until it is shaken vigorously after which, because the viscosity has decreased, the liquid flows easily. The same effect is used in a more subtle way in ball-point pens which contain shear-thinning ink. The viscosity is decreased when the ball rotates thereby allowing the ink to flow onto the paper. High viscosity is recovered and flow stops (usually) when the ball is stationary so that a ball-point pen in the pocket doesn't leak. These are examples of shear-thinning behaviour.

In the case of salad cream and ketchup, another effect called *thixotropy* is also present, this being the term used to describe fluids which take time to adjust to the state of shear. If it were not for this, once the shaking ceased the fluid would return to its high-viscosity state. In addition to being shear thinning, many water-soluble polymers have the effect of reducing the resistance to fluid motion, even in concentrations so low (parts per million) that there is no measurable change in the viscosity of the base solvent. This effect, called *drag reduction*, is still not understood, but may be associated with viscoelasticity, another property of some non-Newtonian liquids.

A few non-Newtonian fluids are shear thickening (or *dilatant*), which is to say that the viscosity increases with shear rate. Starch-based liquids, such as custard made from powder (e.g. Birds custard powder), are shear thickening. The surface of a thick paste of custard powder and milk appears to be almost solid if stabbed with a spoon, though the spoon will sink gently into the paste due to its own weight. If the spoon is moved rapidly through the paste the liquid surface appears to fracture, but then gently flows back together. This is shear-thickening behaviour.

Some materials, such as gels, lubricating greases, ice cream and margarine, appear to be solid when subjected to low shear stresses: unconfined they maintain their shape without deformation due to gravity but become fluid once the shear stress exceeds a certain threshold value, called the yield stress.

Typical variations in shear stress and viscosity with shear rate are shown in Figure 2.6 for the four basic fluid types:

- Newtonian   – μ = constant independent of shear rate
- shear thinning   – μ decreases with shear rate

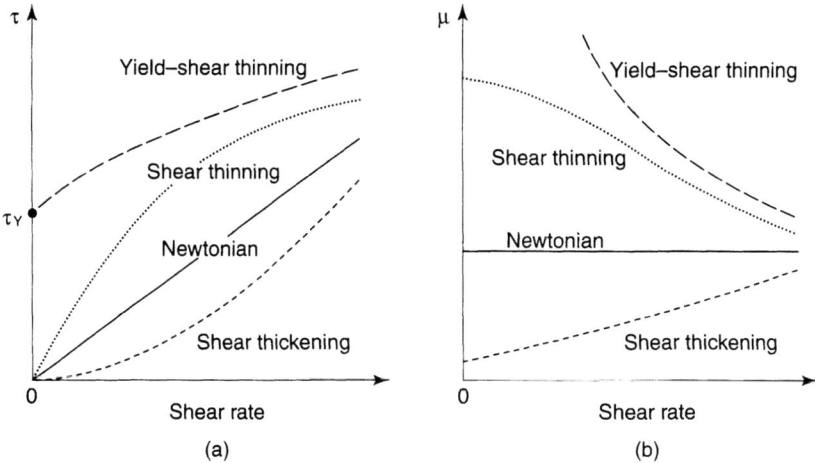

Figure 2.6 *Variation of (a) shear stress and (b) dynamic viscosity with shear rate for non-Newtonian liquids*

- shear thickening   – μ increases with shear rate
- yield stress   – $\tau <$ yield stress, shear rate = 0 (i.e. no flow, material is 'solid')
  – $\tau >$ yield stress, shear rate > 0 (i.e. flow occurs, initially at least shear thinning).

The flow behaviour of non-Newtonian fluids is complicated and still far from understood, particularly for those which are elastic or thixotropic. It is fortunate for us that all gases and many liquids of engineering interest, especially water, are Newtonian in character. But we should not forget that in many respects Newtonian fluids are the exception rather than the rule.

2.10 **Bulk modulus of elasticity and compressibility**

We have already seen that viscosity is the property which relates the rate of change of shape of a fluid to shear stress. The corresponding property which *relates the change in volume to a change in pressure* is the *bulk modulus of elasticity*, K, i.e. K is the property which characterises the *compressibility* of a fluid. Although all fluids are compressible to some extent, gases are far more so than liquids and we often distinguish between them primarily on this basis. In fact, as we discuss in Chapter 7, in many practical gas-flow problems the variation in pressure is sufficiently small that we can treat the gas as though it were incompressible.

The bulk modulus of elasticity is defined by the following equation:

$$K = -\frac{\delta p}{\delta \vartheta / \vartheta} = -\frac{\vartheta \; \delta p}{\delta \vartheta}$$

where $\delta \vartheta$ is the change in the fluid volume ϑ due to a pressure change δp. Since an increase in pressure (i.e. $\delta p > 0$) causes a decrease in volume (i.e. $\delta \vartheta < 0$), a minus sign is introduced to ensure that K is a positive quantity. In the limit of an infinitesimal change in pressure which produces an infinitesimal change in volume, we can write

$$K = -\vartheta \; \frac{dp}{d\vartheta}$$

where $dp/d\vartheta$ is the derivative or gradient of the $p - \vartheta$ curve during a compression (or expansion) process. If we consider a fixed mass of fluid m, then

$$m = \rho \vartheta$$

and, since there is no change in mass during compression or expansion,

$$dm = \vartheta d\rho + \rho d\vartheta = 0$$

so that

$$\frac{d\vartheta}{\vartheta} = -\frac{d\rho}{\rho}$$

and finally

$$K = \rho \; \frac{dp}{d\rho}.$$

This is a far more satisfactory definition of K than that involving ϑ because, as should be evident, like the density ρ, the bulk modulus of elasticity is a thermodynamic property. The *inverse of K* (i.e. $1/K$) is called the *compressibility*.

In general, compression and expansion are thermodynamic processes involving an increase or decrease in the pressure accompanied by a corresponding change in the temperature of a fluid. To define completely a property which quantifies the compressibility of a fluid, therefore, we need to specify either the temperature or the thermodynamic process itself. A *process in which the temperature remains constant is called isothermal* and one in which there is no heat transfer is called *adiabatic*. If there is *neither heat transfer nor friction* the *process* may be considered to be *isentropic* (meaning that the thermodynamic property known as *entropy* remains constant). The distinction between these two processes is important for gases but not for liquids.

Values of K for a range of commonly encountered liquids are listed in Table A.4. For an ideal gas, the pressure–density relationship for an isentropic process can be shown to be

$$p = \text{constant} \times \rho^\gamma$$

where γ is the ratio of specific heats C_p/C_v (see section 2.5). We then have

$$\frac{dp}{d\rho} = \gamma \times \text{constant} \times \rho^{\gamma-1}$$

$$= \frac{\gamma p}{\rho}$$

and the *isentropic bulk modulus of elasticity* K_s (the subscript 's' indicates 'isentropic') is

$$K_s = \gamma p.$$

2.11 **Speed of sound**

Sound travels through a fluid in the form of low-amplitude pressure fluctuations or waves – a process which can be regarded as isentropic. The speed at which such waves propagate through a fluid is called the speed of sound c and can be shown to be given by

$$c^2 = \left.\frac{dp}{d\rho}\right|_s = \frac{K_s}{\rho}.$$

In section 2.5 we stated that the equation of state for an ideal gas is

$$p = \rho RT \tag{2.6}$$

and we have just shown that for an isentropic process

$$K_s = \gamma p.$$

If we combine the three equations above we have

$$c = \sqrt{(\gamma RT)}$$

i.e. since the ratio of specific heats γ and the gas constant R are constants for a given fluid, we see that for an ideal gas the speed of sound c is proportional to the square root of the absolute temperature T. Thus the speed of sound is lower on a cold day than on a hot one: between early morning and early afternoon the air temperature in the Black Rock Desert in Nevada might increase from 0°C to 40°C. The corresponding increase in the speed of sound is from 337 m/s to 355 m/s, which explains why it was advantageous for the successful attempt of the Project Thrust Supersonic Car to break through the sound barrier to take place early in the day (the actual air temperature was reported to be between 5 and 8°C). Also, since

$$R = \frac{\Re}{M}$$

we have

$$c = \sqrt{\frac{\gamma \Re T}{M}} \tag{2.7}$$

and we see that the speed of sound will be much higher for low molecular weight gases such as hydrogen and helium than for heavier gases such as air and carbon dioxide. Values for the speed of sound for air and a range of other gases at 20°C are given in Table A.5, although clearly this was not strictly necessary since the speed of sound is not an independent property but can be calculated at any temperature from equation (2.7) if γ and M are known.

Values of the isentropic modulus for liquids (see Table A.4) are very high compared with the values of K_s for most gases, so that the speed of sound for liquids is also generally higher except for gases with low molecular weight.

EXAMPLE 2.7

Calculate the speed of sound for air ($\gamma = 1.4$, $M = 29$), helium ($\gamma = 1.63$, $M = 4$) and water ($\rho = 998$ kg/m^3, $K = 2.19 \times 10^9$ Pa) at 20°C. The universal gas constant $\Re$ has the value 8314.5 m^2/s^2.K.

SOLUTION

$T = 20 + 273 = 293$ K; $\quad \Re = 8314.5$ m^2/s^2.K

For air: $\gamma = 1.4$, $M = 29$, so that

$$c = \sqrt{\frac{1.4 \times 8314.5 \times 293}{29}} = 343 \text{ m/s}.$$

For helium: $\gamma = 1.63$, $M = 4$, so that

$$c = \sqrt{\frac{1.63 \times 8314.5 \times 293}{4}} = 996 \text{ m/s}.$$

For water, $K = 2.19 \times 10^9$ Pa, $\rho = 998$ kg/m^3, so that

$$c = \sqrt{\frac{2.19 \times 10^9}{998}} = 1481 \text{ m/s}.$$

Comment: Only the value of c for helium is significantly different from the values in Table A.5, and even for helium the difference is only 1.1%.

2.12 Vapour pressure, boiling and cavitation

It is a common observation that liquid in a container open to the atmosphere will evaporate. What we mean by this is that liquid molecules just below the liquid surface have sufficient momentum to overcome the intermolecular cohesive forces we discussed in section 2.1 and escape in vapour form into the atmosphere. If the liquid is placed in a closed container and the space above the liquid surface evacuated (i.e. any air is pumped out and the pressure reduced to zero), evaporation of the liquid will cause the pressure in the container to increase until an equilibrium state is reached when as many molecules leave the liquid surface to create vapour as return to the liquid. Under these equilibrium conditions the vapour is said to be saturated and the pressure is the *saturated vapour pressure* p_V, usually referred to simply as the *vapour pressure*. The corresponding temperature is called the *saturation temperature*. Since molecular activity increases with temperature, the vapour pressure also increases with temperature. The variation of vapour pressure with temperature for water is given in Table A.2 and shown in graphical form in Figure 2.7.

If the pressure within a body of liquid equals the vapour pressure corresponding to the liquid temperature, vapour bubbles are produced within the liquid until all the liquid has become vapour. This is the process we call *boiling*. In a closed container, the production of vapour increases the pressure until equilibrium conditions are reached corresponding to a point on the $p_V(T)$ curve. As we shall show in Chapters 7 and 8, the pressure in a fluid stream decreases if the fluid velocity increases, for example, in flowing through a valve or nozzle. If the pressure within the liquid stream falls below the vapour pressure corresponding to the liquid temperature, internal boiling will begin an undesirable phenomenon called *cavitation* (see section 8.10 of Chapter 8). Since the vapour pressure increases with temperature, the danger of cavitation also increases, for example, in poorly designed domestic-heating systems.

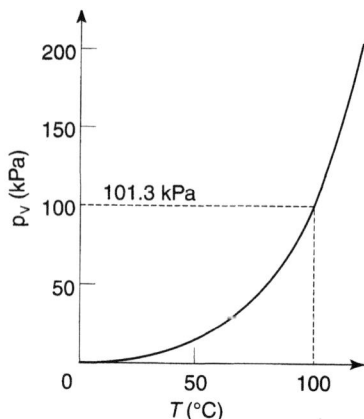

Figure 2.7 *Variation of saturated vapour pressure with temperature for water*

EXAMPLE 2.8
The atmospheric pressure at the top of Mount Everest (height 8848 m) is
31 kPa. At what temperature does water boil at this altitude? What pressure
would be required for the boiling point to be raised to 100°C?

SOLUTION
$p_1 = 31$ kPa; $T_2 = 100$°C
From Table A.2 we have

$$T_1 = 70°C \quad \text{and} \quad p_2 = 1.013 \times 10^5 \text{ Pa.}$$

Comment: 100°C corresponds to what we normally think of as the boiling
point of water. Due to the low pressure at high altitude, the boiling point
is reduced by 30°C. The pressure has to be increased to 1 atm, or
1.013×10^5 Pa, for the boiling point to return to 100°C.

2.13 Surface tension and contact angle

The discussion of the continuum hypothesis in section 2.3 was limited to situations in which molecules of fluid in the interior of a fluid interacted with molecules of the same fluid in the same thermodynamic state. As we have just seen,
at the surface of a liquid in contact with either a gas or its own vapour, we need
to take into account the fact that molecules constantly cross the surface. A more
general term for the surface which separates two fluids, such as a liquid and a gas
or two immiscible liquids, such as water and mercury or oil and water, is

interface. For liquid in a tube, the interface is also called a *meniscus*. Although the chemistry and physics are complex, for most practical purposes such an interface can be treated as a skin or membrane in tension and this leads to the identification of a new fluid property called *surface tension*, σ. Surface tension for a pure liquid decreases almost linearly with increasing temperature and also depends upon whether the liquid is in contact with its own vapour or with air (or some other gas). An important but difficult to quantify influence on surface tension is contamination of the liquid, either due to unwanted impurities or detergents which markedly decrease σ. The values of surface tension listed in Table A.4 are for pure liquid.

Figure 2.8 shows a curved line drawn in the free surface of a liquid. The arrows which are everywhere normal to the curve and tangential to the surface represent the force due to surface tension which is defined as *the tensile force per unit of length of surface* and has the units N/m.

The formation of liquid droplets and of soap bubbles can be explained using the concept of surface tension. Consider the circular segment of a spherical surface of radius R shown in Figure 2.9. If the pressure difference between the inside of the surface and the outside is Δp, then for the segment to be in static equilibrium requires

$$2\pi r \sigma \sin \theta - \pi r^2 \Delta p = 0.$$

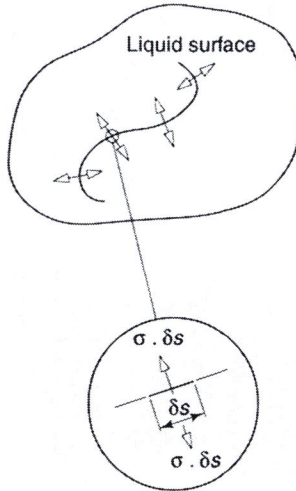

Figure 2.8 *Plan view of a surface to illustrate surface tension*

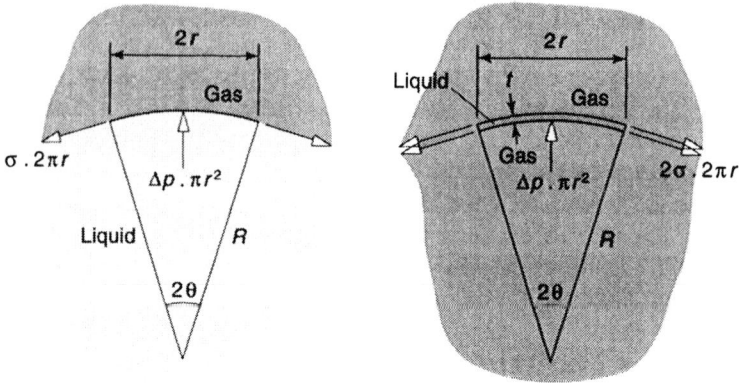

Figure 2.9 *Forces acting on (a) liquid droplet (b) soap bubble*

From the geometry we have

$$\sin \theta = \frac{r}{R},$$

so that the static equilibrium equation reduces to

$$\Delta p = \frac{2\sigma}{R}. \qquad (2.8)$$

If we apply this result to a spherical mass of liquid, it shows that the internal pressure increases as the drop size decreases. In fact, on earth equation (2.8) is limited to small droplets because, as R increases, gravitational effects (which we ignored in deriving equation (2.8)), first distort the drop shape and eventually create pressure differences too great to be withstood by the surface tension force.[*] In the zero-gravity conditions of outer space no such limitation exists and spherical liquid drops of any size are theoretically possible.

In the case of a bubble surrounded by gas, the equivalent to equation (2.8) is

$$\Delta p = \frac{4\sigma}{R},$$

the increase in Δp by a factor of 2, compared with the result for a liquid drop, being a consequence of the fact that a bubble has both an inner and an outer surface of effectively the same radius R. Because the gas which fills a bubble has a much lower density than the liquid which forms its surface, bubbles can reach

---

[*] The variation of pressure in fluid at rest is discussed in Chapter 4.

much larger sizes than drops before gravitational effects have a significant influence.

EXAMPLE 2.9
The surface tension for petrol is 2.16×10^{-2} N/m. Calculate the pressure inside a petrol droplet 2 μm in diameter created by a fuel injection nozzle if the external pressure is 2.5 bar.

SOLUTION
$\sigma = 2.16 \times 10^{-2}$ N/m; $R = 10^{-6}$ m; $p_E = 2.5 \times 10^5$ Pa
From equation (2.8)

$$\Delta p = p_I - p_E = \frac{2 \times 2.16 \times 10^{-2}}{10^{-6}}$$

$$= 4.32 \times 10^4 \text{ Pa}$$

so

$$p_I = (2.5 \times 10^5) + (4.32 \times 10^4) = 2.93 \times 10^5 \text{ Pa}.$$

We started this chapter with the word 'wet'. Whether or not a solid surface is wetted by a liquid depends upon the extent to which there is an attraction between the liquid molecules and the surface. The degree of attraction is measured by the angle, called the *contact angle* θ, at which the liquid meets the surface, a quantity which depends upon the same factors as surface tension and, in addition, upon the nature of the surface:

$$\theta < 90° \quad \text{wetting}$$
$$\theta > 90° \quad \text{non-wetting}.$$

Water on a clean, grease-free glass surface has a contact angle practically equal

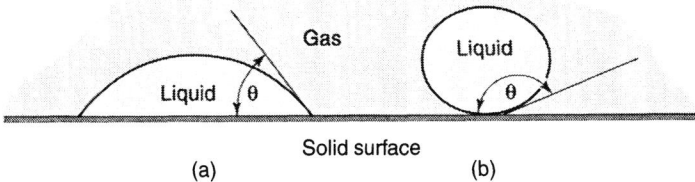

Figure 2.10 *Drops of (a) a wetting and (b) a non-wetting liquid on a horizontal surface*

to zero while for mercury the value is about 130°. The combined effects of surface tension and contact angle thus determine the shape of a liquid drop on a horizontal surface, as shown in Figure 2.10. As we shall show in section 4.8 of Chapter 4, a wetting fluid is drawn into a vertical small-diameter tube due to surface tension whereas the surface of a non-wetting liquid is depressed. This effect is known as *capillarity* and for large-diameter tubes is evident in curvature of the liquid surface very close to the tube wall.

2.14 Summary

In this chapter we have shown that the differences between solids, liquids and gases have to be explained at the level of the molecular structure. The continuum hypothesis allows us to characterise any fluid and ultimately analyse its response to pressure difference and shear stress through macroscopic physical properties which can be defined at any point in a fluid, and which are dependent only upon temperature and pressure. The most important of these properties are density and viscosity, while certain problems are also influenced by bulk modulus of elasticity, vapour pressure, surface tension and contact angle. We showed that the bulk modulus of elasticity is a measure of fluid compressibility and determines the speed at which sound propagates through a fluid. We also introduced the ideal-gas law and derived an equation for the soundspeed.

The student should be able to

- state what is meant by the continuum hypothesis
- calculate the number of molecules in a given volume of any fluid
- define fluid density as

$$\rho = \frac{m}{\vartheta}$$

- define dynamic viscosity as

$$\mu = \frac{\tau}{\text{shear rate}}$$

- calculate the soundspeed for any fluid from

$$c^2 = \frac{K_s}{\rho}$$

and for an ideal gas from

$$c^2 = \frac{\gamma \Re T}{M}$$

- state what is meant by saturated vapour pressure
- define surface tension as

$$\sigma = \text{interfacial force per unit length}$$

- use the tables in the appendix to look up values for fluid properties and use them in calculations.

2.15 Self-assessment problems

2.1   Calculate the sizes of cubes which contain one billion (10^9) molecules of the following substances at STP: air, water, hydrogen, Freon 12 and hydrogen sulphide (H_2S).
(Answers: 3.42 μm, 0.31 μm, 3.42 μm, 0.53 μm, 3.42 μm)

2.2   Calculate the relative density and specific weight of air at 500°C, 1 atm and of methanol at 20°C, 1 atm.
(Answers: 0.379, 4.48 N/m³; 0.792, 7760 N/m³)

2.3   Calculate the specific gas constant, the density and the speed of sound for sulphur hexafluoride (an insulating gas used in high-voltage circuit breakers) at 100°C and 3 bar. Take the molecular weight of SF_6 as 146 and the ratio of specific heats as 1.085.
(Answers: 56.9 m²/s².K, 14.1 kg/m³, 152 m/s)

2.4   Calculate the density of water at a depth of 5000 m where the pressure is about 5000 bar.
(Answer: 1150 kg/m³)

2.5   A concentric-cylinder viscometer has the dimensions $R = 25$ mm, $H = 80$ mm and $\delta = 0.15$ mm. Calculate the shear stress and the torque exerted on the inner cylinder by ethylene glycol at STP if the rotation speed is 120 rpm.
(Answers: 41.7 Pa, 0.013 N.m)

2.6   The viscosity of a non-Newtonian liquid is 0.5 Pa.s for a shear rate du/dy of 10 s⁻¹. At very high shear rates the viscosity falls to a constant value of 0.2 Pa.s. Assuming the shear stress τ obeys the equation

$$\tau = \tau_Y + C\frac{du}{dy}$$

where C is a constant, calculate the value of the yield stress τ_Y.
(Answer: 3 Pa)

2.7   Calculate the speed of sound for petrol at STP.
(Answer: 1187 m/s)

2.8   An experiment is being designed in which water has to boil at the normal
      body temperature of 37°C. What pressure is required?
      (Answer: 6.44 kPa)

2.9   The mass of liquid used to create a soap bubble 100 mm in diameter is
      0.1 mg. The surface tension of the liquid is 0.03 N/m, its density is
      1000 kg/m^3 and its molecular weight is 18. Calculate the pressure
      difference between the inside and the outside of the bubble and the
      thickness of the soap film. Is the continuum hypothesis satisfied?
      (Answers: 2.4 Pa, 3.2 nm, just)

CHAPTER 3

Units of measurement, dimensions and dimensional analysis

This chapter is about dimensions and how they can help us in the analysis of physical problems. Although the illustrative examples are usually limited mainly to fluid flow, it is important to realise that the basic principles introduced here apply to any branch of physics or engineering. At first sight dimensional analysis can appear to be abstract and mystifying but actually involves little more than simple arithmetic and the basic principle that each term in any equation or function, involving physical quantities, must have the same overall dimensions.

We start by discussing the units of measurement and dimensions which are essential parts of any physical quantity. We then introduce the units for flow quantities and for the physical properties of fluids associated with the International System of Units (kilogram, metre, second) followed by the corresponding dimensions (Mass, Length, Time) for these quantities. We show how dimensions and units can be multiplied and divided and demonstrate that the overall dimensions of each term in any formula or relationship involving physical quantities have to be the same. We then show how this 'principle of dimensional consistency' leads to a systematic procedure which allows the physical quantities which describe any problem to be combined to produce a smaller number of non-dimensional groups. Some of these non-dimensional groups are especially important and are given names, such as the Reynolds and Mach numbers. The chapter concludes with the topics of dynamic similarity and scaling which allows us, for example, to predict the aerodynamic performance of a full-scale racing car from a small-scale wind tunnel test.

3.1 **Units of measurement**

The measure or value of any physical quantity, such as acceleration, force, pressure, density or viscosity, is meaningless unless its units are also stated. All physical quantities have units and also dimensions – the two always go together. One of the few exceptions to this general rule is the plane angle (measured in radians) which can be thought of as the ratio of two lengths (e.g. $\pi =$ circumference/diameter or, more generally, plane angle = arc length/radius) and as a consequence has neither units nor dimensions (*the dimension of any* so-called *non-dimensional quantity is* 1).

In ancient times basic units of length were often based upon the size of parts of the human anatomy, such as the hand, forearm (cubit) or foot, and extended by multiplying factors to suit particular applications, e.g. yard (3 feet), chain (22 yards), furlong (10 chains) and mile (8 furlongs). Over the centuries a wide array of units has evolved, particularly for length and weight and closely associated quantities such as area and volume: inches, metres, pounds, hundredweight, tons, grams, kilograms, poundals, slugs, acres, hectare, pints, gallons, pecks, bushels, etc. With the addition of seconds, minutes, hours and other units of time, together with units of temperature (degrees Celsius or centigrade, Fahrenheit and kelvin), these units of measurement are sufficient to describe all the physical quantitics we shall consider in this book, and indeed all we are likely to encounter in much of engineering and physics but not including electrical quantities. As is the case with pounds, shillings, pence, yen, krone, francs, marks and other denominations of currency, for limited daily use the preference for one system of units over another is essentially a matter of history and familiarity.

The units of most of the quantities we encounter in engineering and science cannot be expressed in terms of units of length or mass or time alone. Instead they have to be expressed in terms of a combination of these basic units or in terms of new units introduced for each quantity. Without a very simple underlying system of units, the conversion between units becomes complicated and prone to error.

EXAMPLE 3.1

What tractive force F lb$_f$ is required to accelerate a car of mass m tons from rest with constant acceleration to a speed V miles per hour in a time t minutes?

SOLUTION

We recognise that this problem requires the use of Newton's second law of motion $F = ma$. In this case the acceleration

$$a = \frac{V}{t}$$

so

$$F = \frac{mV}{t}$$

and the resulting units of force must be tons × (miles ÷ hours) ÷ minutes (we shall return to how we multiply and divide units in the next section).

It is quite clearly not very useful to have a force expressed in such peculiar units as ton mile/hr min so we need to introduce conversion factors in the hope of producing a more familiar unit of force: 1 (long) ton = 2240 lb_m, 1 mile = 5280 ft, 1 hour = 3600 s, 1 minute = 60 s. The result is then

$$F = 2240\,m \times \frac{5280\,V}{3600} \times \frac{1}{60\,t} = \frac{54.76\,mV}{t}\,\frac{lb_m.ft}{s^2}$$

which is not a great deal more helpful.

We now recognise that the weight W of a mass m is given by $W = mg$, where g is the gravitational acceleration (approximately 32.2 ft/s²), so that 1 $lb_m \times$ 32.2 ft/s² must be equivalent to 1 lb_f and our force

$$F = \frac{54.76\,mV}{32.2\,t} = \frac{1.702\,mV}{t}\,lb_f.$$

If this example doesn't convince the reader that a more coherent system of units would be preferable, nothing will!

3.2 International System of Units (SI)

The units of measurement now preferred for most engineering purposes are those of the International System of Units (Systeme International d'Unités or, simply, SI) which is a coherent system based upon the *kilogram* (kg) for mass, *metre* (m) for length, *second* (s) for time, *kelvin* (K) for absolute temperature and (though of no relevance here) *ampere* (A) for current. Although the name was not adopted until 1960, the SI has its origin in a system adopted by the National Assembly of France in 1795 and has been in general use throughout continental Europe for many years. Within the UK the use of the SI has only recently become widespread and it may come as something of a surprise to learn that its use throughout the United States was legalised by the US Congress as long ago as 1866. The relevant British Standard, in which all SI units are defined and recommendations given for their use, is BS 5555: 1993 (identical with ISO 1000: 1992).

3.3 Dimensions

While there is an almost unlimited choice of units of measurement, the same is not true of dimensions which are far more basic in character. For our present purposes we need only *mass* (dimensional symbol M), *length* (L) and *time* (T). Should we wish to extend consideration to thermodynamics and heat transfer, we would need to introduce a *dimension for temperature* (θ).

The symbols, units and dimensions of all the physical quantities we shall encounter in this book are listed in Table A.6. The symbols are not just those used throughout this book but to a large extent those used in engineering fluid mechanics generally. To assist the reader, the word form of each of the Greek symbols has also been included in the table. In addition to the three *base or primary units* (kg, m, s), there are a number of *derived units* which are particular combinations of kg, m and s and which are given special names, including *hertz* (Hz), *newton* (N), *pascal* (Pa), *joule* (J) and *watt* (W). Although obviously named after great scientists, these derived units are never capitalised when written out in full. To avoid very large or small numbers with many zeros, use of SI also involves prefixes attached to units which correspond to multiplying factors which are powers of 1000. These prefixes are listed in Table A.7. Since it is difficult to remember the dimensions of quantities such as dynamic viscosity (M/LT) and power (ML^2/T^3), it is well worth remembering that there is a *one-to-one correspondence between the basic SI units (i.e. kg, m and s) and the dimensions M, L and T,* so the units of any quantity, which may appear in thermodynamic and other property tables, can always be used to work out its dimensions. The reverse is also true but less useful.

EXAMPLE 3.2

Convert the derived SI units for pressure, dynamic viscosity and power to basic SI units and hence find the dimensions of these three quantities.

SOLUTION

The derived SI unit for pressure

$$Pa = \frac{N}{m^2} = \frac{kg.m}{s^2} \cdot \frac{1}{m^2} = \frac{kg}{m.s^2}.$$

Thus

$$[p] = \frac{M}{LT^2}.$$

Similarly, for dynamic viscosity, the derived unit is Pa.s, so that we have

$$Pa.s = \frac{kg}{m.s^2} \cdot s = \frac{kg}{m.s}$$

and so

$$[\mu] = \frac{M}{LT}.$$

Finally for power we have

$$W = \frac{J}{s} = \frac{N.m}{s} = \frac{kg.m}{s^2} \cdot \frac{m}{s} = \frac{kg.m^2}{s^3}$$

and so

$$[P] = \frac{ML^2}{T^3} .$$

In this example we have introduced the convention of placing *square brackets [ ]* *around a quantity* to *denote equations which refer only to the dimensions of* *physical quantities.*

One quantity which frequently causes difficulties for students in working out dimensional problems is rotational speed N rpm (rpm is a commonly used non-SI unit). Since one complete revolution corresponds to 360° or 2π rad and there are 60 seconds in a minute, it should be clear that the angular velocity ω in rad/s which corresponds to N rpm is given by $\omega = 2\pi N/60$. Since 2, π and 60 are pure numbers, and so non-dimensional, the dimensions of both ω and N must be $1/T$. It is a common mistake to assume that angular velocity is no different dimensionally from linear velocity. To emphasise the point, recall that the linear velocity V of a point on the circumference of a wheel of radius R rotating at angular velocity ω is given by $V = \omega R$, from which we have $\omega = V/R$ and so

$$[\omega] = \left[\frac{V}{R}\right] = \frac{L}{T} \times \frac{1}{L} = \frac{1}{T} ,$$

as before.

3.4 Arithmetic of combining dimensions and of combining units of measurement

If two physical quantities are combined either by multiplication or division, then the dimensions of the resulting quantity are obtained from the dimensions of the original quantities by precisely the same arithmetic process. As with normal arithmetic, we can cancel dimensions, multiply powers of them together by adding indices, and divide by subtracting indices. It should be self-evident that *dimensions can be neither added nor subtracted*: it makes no sense, for example, to add Length to Time. All of the foregoing applies equally to the manipulation of units.

EXAMPLE 3.3

Work out the SI units of a quantity resulting from multiplying 12.3×10^9 kg by 67.9 μm and dividing by 17 ms².

SOLUTION

$$\frac{12.3 \times 10^9 \text{ kg} \times 67.9 \text{ μm}}{17 \text{ ms}^2} = \frac{12.3 \times 10^9 \text{ kg} \times 67.9 \times 10^{-6} \text{ m}}{17 \times (10^{-3})^2 \text{ s}^2}$$

$$= 49.13 \times 10^9 \frac{\text{kg.m}}{\text{s}^2} = 49.13 \text{ GN}.$$

An important point with regard to combinations of SI units is that it is essential to separate the individual units either by a dot, as here, or by a space to avoid confusion between, say, ms meaning millisecond and m.s (or m s) for metre second.

EXAMPLE 3.4

What are the SI units and dimensions of the quantity which results from dividing an acceleration by a velocity?

SOLUTION

Acceleration a has SI units m/s² and velocity V the units m/s. The units of $a \div V$ are thus

$$\frac{\text{m}}{\text{s}^2} \div \frac{\text{m}}{\text{s}} = \frac{\text{m}}{\text{s}^2} \times \frac{\text{s}}{\text{m}} = \frac{1}{\text{s}} \text{ or s}^{-1}, \text{ i.e. the units of frequency.}$$

As mentioned above, the usual way of stating the dimensions of a physical quantity is to use square brackets, e.g.

$$[a] = L/T^2 \quad \text{and} \quad [V] = L/T.$$

The dimensions of a/V are then found as follows

$$\left[\frac{a}{V}\right] = \frac{L}{T^2} \times \frac{T}{L} = \frac{1}{T},$$

i.e. the dimension of frequency, as we should expect since we found the unit to be s⁻¹.

Note that it is generally less confusing and safer to combine groups of units or dimensions by multiplication rather than division: in this instance we inverted the dimensions of V to change from $L/T^2 \div L/T$ to $L/T^2 \times T/L$.

EXAMPLE 3.5

What are the units and dimensions of the quantity $p/\rho V^2$, where p is pressure, ρ is density and V is velocity?

SOLUTION

We proceed as before, first writing down the dimensions and units of p, ρ and V and then determining the dimensions and units of $p/\rho V^2$:

$$\text{dimensions:} \quad [p] = \frac{M}{LT^2}; \quad [\rho] = \frac{M}{L^3}; \quad [V] = \frac{L}{T}$$

units: Pa kg/m^3 m/s.

Then

$$\text{dimensions:} \quad \left[\frac{p}{\rho V^2}\right] = \frac{M}{LT^2} \times \frac{L^3}{M} \times \left(\frac{T}{L}\right)^2 = 1$$

$$\text{units:} \quad \frac{kg}{m.s^2} \times \frac{m^3}{kg} \times \frac{s^2}{m^2} = 1$$

Note that in determining the units of $p/\rho V^2$ we have made use of the fact that the derived SI units Pa and N are defined as N/m^2 and $kg.m/s^2$ respectively.

The end result here is highly significant: $p/\rho V^2$ has both dimensions and units of unity (not zero) which tells us that $p/\rho V^2$ is non-dimensional (the word dimensionless is also often used). A consequence is that the value of $p/\rho V^2$ is independent of the units used for p, ρ and V provided we work with a consistent system of units. The same is true for any non-dimensional group of quantities. We can also conclude that the dimensions and units of any number (e.g. $\pi = 3.1415927...$) are unity.

EXAMPLE 3.6

Suppose $p = 7$ bar, $\rho = 2$ kg/m^3 and $V = 10$ m/s. Calculate the value of $p/\rho V^2$ in both SI and British units.

SOLUTION

In consistent SI units, the value of $p/\rho V^2$ is

$$\frac{p}{\rho V^2} = \frac{7 \times 10^5}{2 \times 10^2} = 3.5 \times 10^3$$

The only conversion needed was for p since the unit of bar ($= 10^5$ Pa), though in common use, is not a primary SI unit.

We now repeat the exercise for p in psi (i.e. lb_f/in^2), ρ in lb_m/ft^3 and V in ft/s.

We note first that 7 bar = 101.2 psi, 2 kg/m^3 = 0.125 lb_m/ft^3 and 10 m/s = 32.81 ft/s (the reader should check each of these conversions). Then

$$\frac{p}{\rho V^2} = \frac{101.2 \times 144 \times 32.2}{0.125 \times 32.81^2} = 3.49 \times 10^3.$$

There are several points to note here. First, the factor 144 in^2/ft^2 has to be introduced to convert psi to lb_f/ft^2: even in the British system of units we have to at least try to be consistent! Second, the factor 32.2 $lb_m.ft/lb_f.s^2$ is needed to convert lb_f to $lb_m.ft/s^2$. Third, the final result is not exactly 3.5×10^3 but 3.49×10^3 as a consequence of the accumulation of small errors in each of the conversions from SI to British units.

3.5 Principle of dimensional consistency (or homogeneity)

Each additive term in any physical equation must have the same dimensions. This statement of the principle of dimensional consistency is the basis of dimensional analysis. It is also the case that *each additive term in a physical equation must have the same units.* Although this book is generally concerned only with fluids and fluid flow, it should be apparent that any statement about dimensions or units applies in any branch of physics or engineering.

EXAMPLE 3.7
Show that each of the following equations is dimensionally consistent:

(a) $e = m.c_0^2$ (b) $v = u + at$ (c) $s = ut + \frac{1}{2}at^2$ (d) $p = B + \rho gh$

(e) $T = 2\pi\sqrt{(l/g)}$ (f) $D = 6\pi\mu VR + \frac{9}{4}\pi\rho V^2 R^2$ (g) $\dot{Q} = \dfrac{\pi R^4 \Delta p}{8\mu L}$

SOLUTION
(a) We recognise $e = mc_0^2$ as the result of Einstein's theory of relativity in which e is the energy released from an object if its mass reduces by an amount m and c_0 is the speed of light. Since

$$[e] = \frac{ML^2}{T^2} \quad \text{and} \quad [mc_0^2] = M \times \left(\frac{L}{T}\right)^2,$$

we are relieved to find that Einstein's equation is dimensionally consistent.

(b) $v = u + at$ is an equation for the velocity v of an object which has initial velocity u and undergoes constant acceleration a for time t. We must consider each of the additive terms separately, as follows:

$$[v] = \frac{L}{T} \quad [u] = \frac{L}{T} \quad [at] = \frac{L}{T^2} \times T = \frac{L}{T}$$

and the equation $v = u + at$ is dimensionally consistent because each additive term has the dimensions of velocity L/T.

(c) $s = ut + \frac{1}{2} at^2$ is the result corresponding to $v = u + at$ for the distance s moved by the object during the time t. In this case:

$$[s] = L \quad [ut] = \frac{L}{T} \times T = L \quad \left[\frac{1}{2} at^2\right] = 1 \times \frac{L}{T^2} \times T^2 = L.$$

Note that the numerical factor $\frac{1}{2}$ has the dimension of unity. We see from this example that no consideration of dimensions can tell us anything about the value of a numerical factor: whatever the numerical value, the dimension is unity.

(d) $p = B + \rho gz$ gives the pressure p at a depth z below the surface of a liquid of density ρ with barometric pressure B at the surface and where g is the acceleration due to gravity. We have

$$[\rho gz] = \frac{M}{L^3} \times \frac{L}{T^2} \times L = \frac{M}{LT^2}$$

which corresponds to the dimensions of p and B, both of which are pressures.

(e) For a pendulum of length l the period of swing T is given by $T = 2\pi\sqrt{(l/g)}$, g again being the acceleration due to gravity. In this case

$$\left[2\pi\sqrt{\frac{l}{g}}\right] = 1 \times 1 \sqrt{\left(L \times \frac{T^2}{L}\right)} = T$$

which corresponds to the dimension of the period T.

(f) $D = 6\pi\mu VR + \frac{9}{4} \pi\rho V^2 R^2$ is Oseen's formula for the drag force D exerted on a sphere of radius R moving with low velocity V through a fluid of density ρ and dynamic viscosity μ. The dimensions of each additive term are:

$$[D] = \frac{ML}{T^2} \quad [6\pi\mu VR] = 1 \times 1 \frac{M}{LT} \times \frac{L}{T} \times L = \frac{ML}{T^2}$$

$$\left[\frac{9}{4}\pi\rho V^2 R^2\right] = 1 \times 1 \times \frac{M}{L^3} \times \left(\frac{L}{T}\right)^2 \times L^2 = \frac{ML}{T^2}$$

Note that one of the terms on the right-hand side of the formula for D includes ρ and the other μ, one is linear in V and the other quadratic, but the equation is still dimensionally consistent.

(g) The final example here involves the Hagen–Poiseuille formula for the pressure drop Δp over a length L for flow of a fluid with dynamic viscosity μ through a tube of radius R at a volumetric flow rate $\dot{Q}$:

$$\dot{Q} = \frac{\pi R^4 \Delta p}{8\mu L}.$$

Since this formula concerns the volumetric flowrate $\dot{Q}$ rather than the mass flowrate (i.e the units are m^3/s not kg/s), we have

$$[\dot{Q}] = \frac{L^3}{T} \quad \text{and} \quad \left[\frac{\pi R^4 \Delta p}{8\mu L}\right] = 1 \times L^4 \times \frac{M}{LT^2} \times 1 \times \frac{LT}{M} \times \frac{1}{L} = \frac{L^3}{T}.$$

Note that the *pressure difference*, Δp, has the *dimensions of pressure*.

3.6 Dimensional versus non-dimensional representation

We illustrate the advantage of converting a dimensional formula to non-dimensional form using the equation of Example 3.7(c)

$$s = ut + \tfrac{1}{2}at^2.$$

We can easily calculate the value of s given appropriate values for u, a and t and so plot s versus t for different combinations of u and a. Even for such a simple formula this is a tedious exercise: for every value of u there is an infinite choice of values for a and even to cover a limited range for u and a we would need to plot a large number of curves. Figure 3.1 shows s versus t for just five values of a with $u = 1$ m/s.

Suppose now we divide each of the three terms in the equation by u^2 and multiply each by a. Then we have

$$\frac{sa}{u^2} = \frac{at}{u} + \frac{1}{2}\frac{a^2t^2}{u^2} = \frac{at}{u} + \frac{1}{2}\left(\frac{at}{u}\right)^2$$

and we observe that the two combined quantities, sa/u^2 and at/u, are both non-

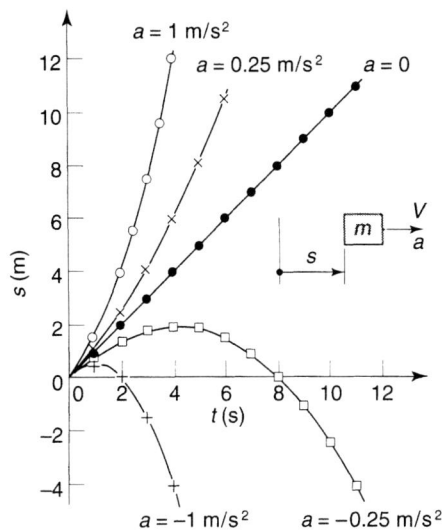

Figure 3.1 *Curves of* s *versus* t *for* s = ut + $\frac{1}{2}$at^2 *with* u = 1 *m/s and* a = 0, ±0.25 *and* ±1.0 *m/s*^{2}

dimensional, i.e.

$$\left[\frac{sa}{u^2}\right] = L \times \frac{L}{T^2} \times \left(\frac{T}{L}\right)^2 = 1 \quad \text{and} \quad \left[\frac{at}{u}\right] = \frac{L}{T^2} \times T \times \frac{T}{L} = 1.$$

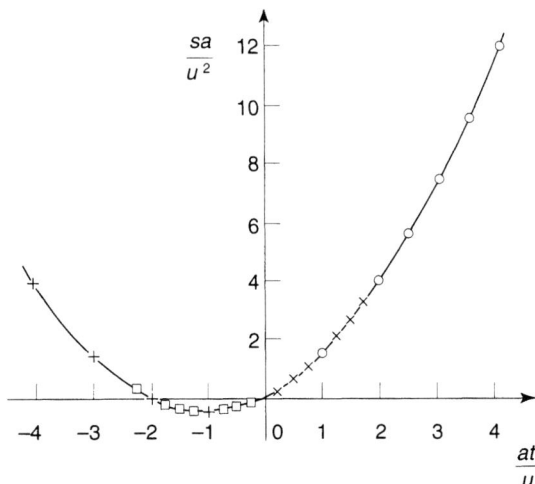

Figure 3.2 *Curve of* sa/u^2 *versus* at/u *corresponding to* s = ut + $\frac{1}{2}$at^2.

Since the two terms on the right-hand side of the non-dimensional equation involve exactly the same combination of variables, at/u, instead of needing to plot s versus t for an infinite number of values for u and an infinite number of values for a, the non-dimensional version of the formula can be represented by a single curve (Figure 3.2) of sa/u^2 versus at/u. It should be obvious that one curve takes a lot less work to produce than five curves. The benefits of non-dimensionalising a more complicated equation are even greater, as are the advantages of plotting experimental data in non-dimensional form. The major benefits of non-dimensional graphs are that information can be presented in a very compact form and, in the case of experimental data, far fewer experiments need be performed. Equally important, in situations where we have only experimental results but little or no theoretical guidance, a non-dimensional plot is more likely to reveal the underlying relationship between all the variables than a dimensional representation.

3.7 **Buckingham's Π theorem**

In the example of section 3.4, we reduced a problem involving four individual dimensional quantities (s, u, a, t) to one involving two non-dimensional groups of quantities $(sa/u^2, at/u)$. In any problem of *dimensional analysis*, as this mathematical process is called, it is obviously convenient to know in advance how many non-dimensional groups will result from a given set of physical quantities. This information is provided by the Π*(pi) theorem* associated with the American physicist Edgar Buckingham (1867–1940):

If a physical process involves n dimensional quantities (or variables) which can be described in terms of j dimensions, then this process can be represented by

$$k = n - j$$

non-dimensional combinations of the dimensional quantities.

Since we are generally limiting ourselves to problems involving physical quantities with dimensions M, L and T, j can only be 1, 2 or 3. We see that the number of non-dimensional quantities (often called groups or numbers) k is always less than the number of dimensional variables n. For example, in the case of $s = ut + \frac{1}{2}at^2$, we see that $n = 4$ (i.e. s, u, a, t) and $j = 2$ (i.e. L, T) and so $k = n - j = 2$, which confirms what we found previously.

Although called Buckingham's Π theorem, the symbol Π, which is the capital version of the Greek letter π, has nothing to do with the constant $\pi = 3.1415927\ldots$ but is simply the symbol chosen by Buckingham to represent a non-dimensional combination of dimensional physical quantities. For our example, we would write

$$\Pi_1 = \frac{sa}{u^2} \qquad \Pi_2 = \frac{ta}{u}.$$

In a more general case the n-dimensional variables reduce to $k = n - j$ non-dimensional combinations of those variables, i.e. $\Pi_1, \Pi_2, \Pi_3, \ldots, \Pi_k$.

3.8 Sequential elimination of dimensions

Although with experience it is often possible to identify the non-dimensional Πs by inspection, it is usually preferable to use a more systematic approach, which is essential for the inexperienced. The method of sequential elimination of dimensions presented here is an essentially foolproof 'recipe' though not the most common procedure in use.*

We illustrate the method by reference once again to the sphere drag part of Example 3.7(f), but pretend now that we know only that D depends upon V, ρ, μ and R but do not know the formula $D = 6\pi\mu VR + \frac{9}{4}\pi\rho V^2 R^2$.

We start by writing

$$D = f(V, \rho, \mu, R), \tag{3.1}$$

which simply means that D is a function of (i.e depends upon or is determined by) V, ρ, μ and R. The quantity D is called the dependent variable while V, ρ, μ and R are the independent variables (i.e. the quantities under our control).

In this case, then, we have five physical variables and $n = 5$. It is vital in any problem of dimensional analysis not to forget the dependent variable, in this case D, when counting the n physical variables!

The dimensions of the physical variables are

$$[D] = \frac{ML}{T^2} \qquad [\rho] = \frac{M}{L^3} \qquad [\mu] = \frac{M}{LT} \qquad [R] = L$$

so we have three dimensions (i.e. M, L, T) and $j = 3$. From Buckingham's theorem we expect to find $k = n - j = 2$ non-dimensional groups.

Our aim is to systematically eliminate the three dimensions M, L and T by multiplying or dividing each of the variables by any one of them (or a power of any one of them). It is important to realise that although *we can start the elimination process with any variable, the end result will always be correct though not always exactly the same.* Suppose we choose to eliminate M first using the

---

* The method is also known as the 'step-by-step approach' following Ipsen (1960).

variable ρ (though we could just as well have chosen μ). Then we have:

$$\left[\frac{D}{\rho}\right] = \frac{ML}{T^2} \times \frac{L^3}{M} = \frac{L^4}{T^2} \qquad \left[\frac{\mu}{\rho}\right] = \frac{M}{LT} \times \frac{L^3}{M} = \frac{L^2}{T}$$

and we can re-write our original equation (3.1) as

$$\frac{D}{\rho} = f_1\left(V, \frac{\mu}{\rho}, R\right)$$

in which we have written $f_1(\ldots)$ to indicate that the dependence of D/ρ on V, μ/ρ and R is not in general the same as the dependence of D on V, ρ, μ and R. At this stage we have reduced the number of variables to four (i.e. D/ρ, V, μ/ρ, R) and the number of dimensions to two (i.e. L, T), so that k is still equal to 2.

We now choose R to eliminate the dimension L from D/ρ, V and μ/ρ, as follows:

$$\left[\frac{D}{\rho R^4}\right] = \frac{L^4}{T^2} \times \frac{1}{L^4} = \frac{1}{T^2} \qquad \left[\frac{V}{R}\right] = \frac{L}{T} \times \frac{1}{L} = \frac{1}{T} \qquad \left[\frac{\mu}{\rho R^2}\right] = \frac{L^2}{T} \times \frac{1}{L^2} = \frac{1}{T}$$

and can write our equation as

$$\frac{D}{\rho R^4} = f_2\left(\frac{V}{R}, \frac{\mu}{\rho R^2}\right)$$

i.e. we now have three variables $(D/\rho R^4, V/R, \mu/\rho R^2)$ and one dimension (T).

Finally we choose V/R to eliminate the dimension T, as follows:

$$\left[\frac{D}{\rho R^4}\left(\frac{R}{V}\right)^2\right] = \frac{1}{T^2} \times T^2 = 1 \qquad \left[\frac{\mu}{\rho R^2}\frac{R}{V}\right] = \frac{1}{T} \times T = 1$$

so that our two non-dimensional groups are $D/\rho V^2 R^2$ and $\mu/\rho VR$, i.e. we can write

$$\Pi_1 = \frac{D}{\rho V^2 R^2} \qquad \Pi_2 = \frac{\mu}{\rho VR}$$

and

$$\Pi_1 = F(\Pi_2) \qquad \text{or} \qquad \frac{D}{\rho V^2 R^2} = F\left(\frac{\mu}{\rho VR}\right).$$

Note that dimensional analysis tells us only that $D/\rho V^2 R^2$ depends upon $\mu/\rho VR$, assuming our original assumption that $D = f(V, \rho, \mu, R)$ was itself correct, but can give us no further information as to the form of the dependence. If we are unable to fully analyse a problem from basic principles, then the non-dimensional representation is of great value in guiding us as to how best to establish a correlation experimentally, not least because it reduces, often significantly, the number of variables we need to deal with independently.

EXAMPLE 3.8
A common method for mixing large batches of liquid food products, plastics, cement and other viscous liquids is with a rotating-paddle mixer. The power P required to rotate the paddle depends upon its rotational speed ω, its radius R, the density of the fluid ρ and its dynamic viscosity μ. Derive a non-dimensional form to represent this dependence.

SOLUTION
Step 1: The functional dependence may be written

$$P = f(\omega, R, \rho, \mu)$$

and so the number of variables $n = 5$.

Step 2: The dimensions of each physical quantity are

$$[P] = \frac{ML^2}{T^3}; \quad [\omega] = \frac{1}{T}; \quad [R] = L; \quad [\rho] = \frac{M}{L^3}; \quad [\mu] = \frac{M}{LT}$$

and the number of dimensions $j = 3$.

Step 3: According to Buckingham's Π theorem

$$k = n - j = 2$$

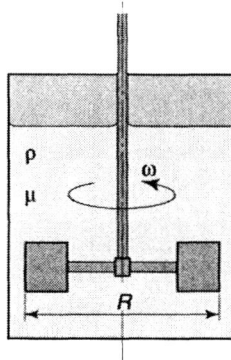

Figure E3.8

and we expect the five dimensional quantities will combine into two non-dimensional groups, Π_1 and Π_2, such that

$$\Pi_1 = F(\Pi_2).$$

Step 4: To find Π_1 and Π_2, we start by selecting ρ to eliminate the dimension M from P and μ:

$$\left[\frac{P}{\rho}\right] = \frac{ML^2}{T^3} \times \frac{L^3}{M} = \frac{L^5}{T^3} \qquad \left[\frac{\mu}{\rho}\right] = \frac{M}{LT} \times \frac{L^3}{M} = \frac{L^2}{T}$$

so that

$$\frac{P}{\rho} = f_1\left(\omega, R, \frac{\mu}{\rho}\right).$$

Step 5: Use R to eliminate L from P/ρ and μ/ρ

$$\left[\frac{P}{\rho R^5}\right] = \frac{L^5}{T^3} \times \frac{1}{L^5} = \frac{1}{T^3} \qquad \left[\frac{\mu}{\rho R^2}\right] = \frac{L^2}{T} \times \frac{1}{L^2} = \frac{1}{T}$$

so that

$$\frac{P}{\rho R^5} = f_2\left(\omega, \frac{\mu}{\rho R^2}\right).$$

Step 6: Use ω to eliminate T from $P/\rho R^5$ and $\mu/\rho R^2$

$$\left[\frac{P}{\rho R^5 \omega^3}\right] = \frac{1}{T^3} \times T^3 = 1 \qquad \left[\frac{\mu}{\rho R^2 \omega}\right] = \frac{1}{T} \times T = 1$$

and finally we have two non-dimensional groups

$$\Pi_1 = \frac{P}{\rho R^5 \omega^3} \qquad \text{and} \qquad \Pi_2 = \frac{\mu}{\rho R^2 \omega}$$

so that

$$\frac{P}{\rho R^5 \omega^3} = F\left(\frac{\mu}{\rho R^2 \omega}\right).$$

Once again this is a perfectly valid result but only one of several possibilities determined by the sequence in which the dimensions M, L and T were eliminated and which physical quantities were chosen to carry out the elimination process.

Alternative non-dimensional groups can be formed by multiplying or dividing the groups, or powers or roots of the groups, which are the 'natural outcome' of dimensional analysis. Some of the other non-dimensional groups we might have obtained in this case are $P\rho^2 R/\mu^3$, $P/\mu\omega^2 R^3$ and $P(\rho^3/\omega\mu^5)^{1/2}$.

3.9 Rayleigh's exponent method

The exponent method of dimensional analysis is attributed to the English physicist Lord Rayleigh (John William Strutt, 1842–1919). This method rests on the rather sophisticated idea that any mathematical function which represents a physical process can be expressed as an infinite power series each term of which, according to the principle of dimensional consistency, must have the same overall dimensions.

We can illustrate the exponent method using the aerodynamic drag example as follows: since

$$D = f(V, \rho, \mu, l)$$

it must be that

$$D = kV^a \rho^b \mu^c l^d + k'V^{a'} \rho^{b'} \mu^{c'} l^{d'} + \cdots$$

where k, k', ..., are numerical constants and a, b, c, d, a', b', c', d', ..., are the exponents (i.e. constants or powers). According to the principle of dimensional consistency, the first term (and all other terms) of the series must have the same dimensions as the dependent variable D, i.e.

$$[D] = [kV^a \rho^b \mu^c l^d]$$

and

$$\frac{ML}{T^2} = 1 \times \left(\frac{L}{T}\right)^a \times \left(\frac{M}{L^3}\right)^b \times \left(\frac{M}{LT}\right)^c \times L^d.$$

The key point is to recognise that since this is a dimensional equation, we require that each dimension balances separately, i.e.

$$L = L^a \times L^{-3b} \times L^{-c} \times L^d = L^{a-3b-c+d}$$

which in turn means that the exponents must balance, i.e.

$$1 = a - 3b - c + d. \tag{3.2}$$

Similarly, from considerations of M,

$$1 = b + c \qquad\qquad (3.3)$$

and for T

$$-2 = -a - c. \qquad\qquad (3.4)$$

As an observation, we note that the number of unknown exponents equals the number of independent quantities (V, ρ, μ, l) while the number of equations equals the number of dimensions (M, L, T). With only three equations we cannot determine all four unknowns: the best we can do is write three of the unknowns in terms of the fourth.

From (3.3) we have

$$b = 1 - c,$$

from (3.4)

$$a = 2 - c,$$

from (3.2)

$$\begin{aligned} d &= 1 - a + 3b + c \\ &= 1 - (2 - c) + 3(1 - c) + c \\ &= 2 - c \end{aligned}$$

and we have now written a, b and d in terms of c.

We now return to the infinite series which can be written as

$$\begin{aligned} D &= kV^{2-c}\rho^{1-c}\mu^{c}l^{2-c} + \cdots \\ &= k(V^2\rho l^2)\left(\frac{\mu}{\rho V l}\right)^{c} + \cdots \end{aligned}$$

In the final expression we have separated the independent variables into those having pure number exponents and those involving the unknown exponent c. We can now divide through by $\rho V^2 l^2$ so that

$$\frac{D}{\rho V^2 l^2} = k\left(\frac{\rho V l}{\mu}\right)^{-c} + \cdots \qquad\qquad (3.5)$$

or

$$\frac{D}{\rho V^2 l^2} = F\left(\frac{\rho V l}{\mu}\right).$$

It is extremely important for the reader to realise that the final result here is not simply

$$\frac{D}{\rho V^2 l^2} = k\left(\frac{\rho V l}{\mu}\right)^{-c}$$

even though this would be a very simple and convenient formula to use once k and c were determined. Unfortunately the exponent method is sometimes presented with the vitally important ' $+\cdots$ ' omitted from equation (3.5) and the unwary reader forgets, or is never told, that he or she is dealing with only one term of an infinite series. Of course, if carried out correctly, the exponent method must produce just the same result as the sequential elimination process.

Although the exponent method of dimensional analysis is probably the method most commonly used, the method of sequential elimination of dimensions is more straightforward and, in the author's opinion, less likely to lead the inexperienced user into difficulty or misunderstanding.

3.10 Inspection method

As stated earlier in this chapter, in many instances it becomes quite straightforward, with experience, to write down the appropriate non-dimensional groups for any given problem essentially from memory or by inspection. For example, if flow velocity V and fluid viscosity μ are involved in a problem, there is a very good chance that the non-dimensional group $\rho V l/\mu$ will appear (obviously the density ρ and a length l are also required). In fact, the inspection approach can be regarded as the method of sequential elimination of dimensions taking one physical quantity at a time. Almost inevitably an approach of this kind is likely to be ad hoc rather than systematic and is not to be recommended for the inexperienced.

3.11 Role of units in dimensional analysis

As we have already pointed out, there is a one-to-one correspondence between the basic SI units, kg, m, s, and the dimensions M, L, T. It should be clear therefore that any problem in dimensional analysis can in principle be worked through using units rather than dimensions, although this is not something this author would recommend. It is, however, extremely valuable to make use of units, which can always be looked up in the tables of thermodynamic data that are usually available during an examination, to determine SI units which are difficult to remember, such as viscosity and power.

EXAMPLE 3.9
The power P required to drive a centrifugal pump, such as that shown in Figure 1.2, depends upon the volumetric flowrate $\dot{Q}$ it is required

to deliver, the pressure rise imposed on the pump Δp, the density of the fluid ρ, the rotation speed of the pump ω, and the impeller radius R. Put the preceding sentence into the form of a non-dimensional equation.

SOLUTION

As always, the first step is to write down the functional dependence

$$P = f(\dot{Q}, \Delta p, \rho, \omega, R)$$

In this case we now write down the units of each of the five quantities involved:

$$P(W); \quad \dot{Q}(m^3/s); \quad \Delta p(Pa); \quad \rho(kg/m^3); \quad \omega(rad/s); \quad R(m)$$

The dimensions of $\dot{Q}$, ρ, ω and R should be obvious, and the next step is to convert the derived units W and Pa to basic SI units:

$$W = \frac{J}{S} = \frac{N.m}{s} = \frac{kg.m}{s^2} \times \frac{m}{s} = \frac{kg.m^2}{s^3}$$

$$Pa = \frac{N}{m^2} = \frac{kg.m}{s^2} \times \frac{1}{m^2} = \frac{kg}{m.s^2}$$

Guided by the units, we can now write down the dimensions of all six physical quantities:

$$[P] = \frac{ML^2}{T^3}; \quad [\dot{Q}] = \frac{L^3}{T}; \quad [\Delta p] = \frac{M}{LT^2}; \quad [\rho] = \frac{M}{L^3}; \quad [\omega] = \frac{1}{T}; \quad [R] = L$$

and can now carry out the dimensional analysis just as before. If we use the sequential elimination method, the steps are as follows:

Use ρ to eliminate the dimension M:

$$\left[\frac{P}{\rho}\right] = \frac{ML^2}{T^3} \times \frac{L^3}{M} = \frac{L^5}{T^3} \qquad \left[\frac{\Delta p}{\rho}\right] = \frac{M}{LT^2} \times \frac{L^3}{M} = \frac{L^2}{T^2}$$

so that

$$\frac{P}{\rho} = f_1\left(\dot{Q}, \frac{\Delta p}{\rho}, \omega, R\right).$$

We now use R to eliminate the dimension L:

$$\left[\frac{P}{\rho R^5}\right] = \frac{L^5}{T^3} \times \frac{1}{L^5} = \frac{1}{T^3}; \quad \left[\frac{\dot{Q}}{R^3}\right] = \frac{L^3}{T} \times \frac{1}{L^3} = \frac{1}{T}; \quad \left[\frac{\Delta p}{\rho R^2}\right] = \frac{L^2}{T^2} \times \frac{1}{L^2} = \frac{1}{T^2}$$

so that

$$\frac{P}{\rho R^5} = f_2\left(\frac{\dot{Q}}{R^3}, \frac{\Delta p}{\rho R^2}, \omega\right).$$

Finally, we use ω to eliminate the dimension T (but note that $[\omega] = 1/T$)

$$\frac{P}{\rho R^5 \omega^3} = F\left(\frac{\dot{Q}}{R^3 \omega}, \frac{\Delta p}{\rho R^2 \omega^2}\right).$$

In this case the six physical quantities have produced three non-dimensional groups as we would expect from Buckingham's Π theorem, since with $n = 6$ and $j = 3$ we have $k = 3$.

3.12 Special non-dimensional groups

What we are typically concerned with in engineering fluid mechanics is the effect of flow velocity V or the rotation speed of a turbomachine (e.g. a turbine, compressor, pump or turbocharger), fluid density ρ, fluid dynamic viscosity μ (or kinematic viscosity v), surface tension σ, soundspeed c and acceleration due to gravity g on the drag force D exerted on an object of size l, or the pressure difference Δp along the length of a pipe of radius R, or the frequency f of periodic disturbances which occur in a fluid as it flows past an object (the list of problems is far from exhaustive). We can write the foregoing as follows:

$$D \text{ or } \Delta p \text{ or } f = F(V, \rho, \mu, \sigma, c, g, l) \tag{3.6}$$

In any engineering problem we have to decide what physical properties of the components and materials involved (fluid or solid) are likely to affect the result, whether the acceleration due to gravity is important, and, if there is motion, what aspects of it need to be included. In the case of fluid flow, we should also consider whether surface finish is important since drag force and pressure change are likely to increase if a surface is rough rather than smooth. What about thermodynamic properties which may be important if there is strong heating or cooling or simply very large pressure changes brought about by the flow itself? Dimensional analysis focuses directly on these questions from the outset. If we include every variable and influence we can think of, problems become intractable (i.e. impossible to solve) and in practice we have to use our judgement to decide which quantities to retain, which to exclude and which might be redundant. For the most part we can only make such decisions on the basis of experience, although common sense also helps. For example, surface tension can be excluded unless free surfaces or interfaces between immiscible fluids (e.g. oil and water) are involved, and there is no point in including v in addition to ρ and μ since $v \equiv \mu/\rho$. As we shall now show, dimensional analysis can also be of great value in guiding our decisions.

It is helpful to ask what physical effect is associated with the independent variables included in the unknown function F(...), equation (3.6). For example, since the momentum of a mass m moving at velocity V is mV and density ρ is mass per unit volume, we can regard the product ρV^2 as a measure of fluid momentum or inertia. Viscosity μ or v will lead to viscous (or shear) forces. Gravitational acceleration g is responsible for pressure increase with depth in a fluid which gives rise to buoyancy forces. Wave motion on a free liquid surface, which results in wave drag, is also a consequence of the acceleration due to gravity while surface tension σ can lead to fine-scale 'capillary' waves on a liquid surface. The soundspeed c is related to the compressibility K of a fluid (i.e. the reduction in volume produced by an increase in pressure) and is associated with the generation and propagation of shock and sound waves through a fluid.

From the list of independent variables included in equation (3.6), the following non-dimensional groups can be formed:

Reynolds number: $Re \equiv \dfrac{\rho V l}{\mu}$ = inertia force : viscous force

Froude number: $Fr \equiv \dfrac{V}{\sqrt{(gl)}}$ = (inertia force : gravity force)$^{1/2}$

Mach number: $M \equiv \dfrac{V}{c}$ = (inertia force : compressibility force)$^{1/2}$

Weber number: $We \equiv \dfrac{\rho V^2 l}{\sigma}$ = inertia force : surface-tension force.

The dependent variables lead to the non-dimensional groups:

Drag coefficient: $C_D \equiv \dfrac{D}{\frac{1}{2}\rho V^2 A}$ = drag force : inertia force

Friction factor: $C_f \equiv \dfrac{\Delta p}{\frac{1}{2}\rho V^2}$ = pressure drop : inertia force

Strouhal number: $St \equiv \dfrac{fl}{V}$ = disturbance period : flow time

Euler number: $Eu \equiv \dfrac{\Delta p}{\rho V^2}$ = pressure force : inertia force.

It is initially confusing to find that there are no absolutes about the definitions of non-dimensional groups. For example, in pipe-flow problems the Reynolds number can be defined using either pipe radius or diameter. The Froude number is sometimes defined as $V^2/(gl)$. Some definitions of C_D and C_f omit the factor $\frac{1}{2}$. These differences in definition are of no consequence provided the definitions used are consistent throughout any particular application.

As we have indicated, each of the non-dimensional groups can be thought of as representing the ratio of two competing effects. The value of each non-dimensional number compared with unity can often be taken as an indicator of what physical quantities are important in a given problem. We consider briefly each number in turn.

Reynolds number ($\rho Vl/\mu$)

In Chapter 2, we identified viscosity μ as the material property of a fluid which distinguishes it from a solid. It is hardly surprising, therefore, that the non-dimensional group which incorporates viscosity, the Reynolds number, plays a role in the majority of flow problems. Osborne Reynolds (1842–1912) was the first professor of engineering at what is now Manchester University. He was the first to demonstrate the phenomenon of cavitation and deduced that the transition from laminar to turbulent flow in pipe flow depended upon the non-dimensional combination of variables which now bears his name. If the Reynolds number is small compared with unity, the fluid inertia is of no importance and there is a balance between viscous and pressure forces. If the Reynolds number is very large much of the flow is dominated by inertia and pressure forces. In the latter case, viscosity still plays a vital role in the region immediately adjacent to any solid surface, known as the boundary layer, where the fluid velocity approaches that of the surface (i.e. zero if the surface is at rest).

Mach number (V/c)

The Austrian physicist and philosopher Ernst Mach (1838–1916) was a pioneer in the field of supersonic aerodynamics. Provided the Mach number is less than about 0.3, a flow can be considered incompressible which is why many gas flows can be treated (with the appropriate values for ρ and μ) in exactly the same way as a liquid flow. For higher values of M, compressibility becomes increasingly important and flow problems are often very difficult to analyse in the transonic region where some regions of a flow are just subsonic ($M < 1$) while others are just supersonic ($M > 1$) and shock waves begin to appear. The abrupt changes in velocity, pressure, temperature and density which occur across a shock wave become increasingly strong as the Mach number becomes increasingly greater than unity. Flows for which $M > 3$ are called hypersonic.

Froude number ($V/\sqrt{(gl)}$)

This parameter is named after William Froude (1810–79), whose name is pronounced to rhyme with food; he was a British naval architect who was interested in free-surface flows and pioneered the use of towing tanks for the study of ship design. Since it can be shown that the speed of propagation of small amplitude waves on the surface of liquid of depth l is $\sqrt{(gl)}$, it can be seen that the Froude number is rather like the Mach number. Free-surface flows with $Fr < 1$ are said to be subcritical (i.e. flowspeed < wavespeed so that small disturbances move faster than the flow) and those with $Fr > 1$ supercritical.

Weber number ($\rho V^2 l/\sigma$)

The German engineer Moritz Weber (1871–1951) is credited with formalising the use of non-dimensional groups as a basis for similarity studies (see section 3.12). Surface-tension effects are only important in fluid flow if the Weber number is of the order of unity or less, which can be the case for small droplets and bubbles, capillary flows and flows of very shallow water. For $We \gg 1$, or if there is no free surface, surface-tension effects are negligible or non-existent.

The remaining four non-dimensional groups arise from dependent rather than independent quantities and so their values represent the outcome of any detailed theoretical analysis or experimental study correlated using these non-dimensional groups.

Strouhal number (fl/V)

Ček Strouhal (1850–1922), the name pronounced as stroohal, was a Czech scientist who, in 1878, carried out experiments on wires 'singing' in the wind, also called aeolian tones. This tendency for *periodic* (i.e. fixed frequency) disturbances to arise in the wake of an object such as a circular cylinder with a completely steady approach flow was explained theoretically by the Hungarian engineer Theodor von Kármán (1881–1963), a major contributor not only to fluid mechanics but to applied mechanics in general. Self-excited oscillations of this type can lead to flow-induced structural vibrations of dangerously high amplitude if the frequency is close to a natural frequency of the structure. The collapse of the Takoma Narrows suspension bridge in Washington State, USA, in 1940 was a consequence of this effect, and the spiral strakes which are wound around tall chimneys are designed to suppress such periodic flow behaviour.

There are situations where a periodic disturbance is imposed on a flow. The oscillation frequency f then becomes an independent variable and the Strouhal number can be used to characterise the resulting flow.

Euler number ($\Delta p / \rho V^2$)

The last of our numbers, named in honour of the Swiss mathematician Leonhard Euler (1707–83) (pronounced oiler), is sometimes written in terms of pressure rather than pressure difference and is sometimes referred to simply as a pressure coefficient, usually with the factor $\frac{1}{2}$ in the denominator, e.g. for representing the pressure variation over the surface of an aerofoil. The other principal application of the Euler number is to liquid flows where the absolute pressure may fall to values comparable with the vapour pressure and lead to internal boiling (at normal ambient temperature) or *cavitation* as it is called. We shall return to the topic of cavitation in Section 8.10.

3.13 **Similarity and scaling**

There would be little point in carrying out experimental studies on scaled-down (or even scaled-up) models if we did not know how the results could be translated (or scaled) to full size. Fortunately this information is just what is provided by dimensional analysis. We require two things:

- *geometric similarity*, which means that the model and full size differ only in size (or scale) but not in shape
- *dynamic similarity* which requires that *each non-dimensional group has the same value for the model and full scale*, i.e. $\Pi_{1M} = \Pi_{1F}$, $\Pi_{2M} = \Pi_{2F}$, etc., where the subscripts M and F refer to the model and full scale, respectively.

We have already seen that a major consequence of dimensional analysis is that the number of separate variables we need to deal with is always reduced by the number of dimensions involved (i.e. usually by three). Although it was not stated at the time, an implicit assumption in dimensional analysis is that we are considering geometrically similar situations. For example, if we write

$$D = \mathrm{f}(V, \rho, \mu, l)$$

for the drag force D exerted on a car of length l, we intuitively realise that if we carry out experiments on a model car it should in principle be a scaled-down replica of the full-size version in all relevant respects. If the model is 1/5 the length of the full size, then the wheels of the model must be 1/5 the diameter of the full size, the width 1/5, etc. The words 'relevant' and 'in principle' were introduced for reasons which relate to both words. Certain aspects of the design play no role in determining the drag (e.g. any aspect of the car's interior) and so are not relevant. Other features may have a minor influence on drag but be too difficult or expensive to reproduce accurately on a model (e.g. gaps between the doors and the body panels, or the trim, or the windscreen wipers).

At first sight the requirement of dynamic similarity may be less obvious than that of geometric similarity and is certainly less straightforward to achieve in practice. In fact, a little thought should reveal that the requirement of dynamic similarity is essentially the same as what is required to reproduce the results of any experiment, i.e. ensure that the independent variables are the same each time. As we have already seen, the power of dimensional analysis is that the number of independent quantities is significantly less than for the problem in dimensional form and it is the non-dimensional combinations of the independent variables which have to be reproduced, not the individual dimensional variables.

In our aerodynamic drag example we have

$$D = f(V, \rho, \mu, l)$$

which tells us that if we carry out an experiment to measure the drag D for given values of V, ρ, μ, l then if we repeat the experiment for exactly the same values of V, ρ, μ, l we shall obtain exactly the same value for D.

In the non-dimensional representation, this equation becomes

$$\frac{D}{\rho V^2 l^2} = F\left(\frac{\rho V l}{\mu}\right)$$

which tells us that for every value of the non-dimensional group of independent physical variables $\rho V l/\mu$ (i.e. the Reynolds number), there will be a corresponding value of the non-dimensional group $D/\rho V^2 l^2$ (i.e. the drag coefficient). The beauty of this is that it is only the values of the two non-dimensional groups $\rho V l/\mu$ and $D/\rho V^2 l^2$ which matter, not the values of their 'constituents'. In other words, we can change the values of V, ρ, μ and l quite freely but if $\rho V l/\mu$ stays the same then $D/\rho V^2 l^2$ will also stay the same.

EXAMPLE 3.10: Case study
A sports car designed for a top speed of 356 kph is being developed for the 24 Heures du Mans endurance race. The prevailing atmospheric conditions are assumed to correspond to an air density of 1.2 kg/m^3 and a dynamic viscosity of 1.8×10^{-5} Pa.s. Calculate the wind-tunnel speed for tests to be carried out on a quarter-scale model car in a pressurised and cooled wind tunnel in which the air density is 4.7 kg/m^3 and the dynamic viscosity is 1.7×10^{-5} Pa.s. If the model test gives a drag force of 1334 N, what is the corresponding drag for the full-size car and the tractive power required, assuming dynamic similarity between the wind-tunnel and full-scale situations?

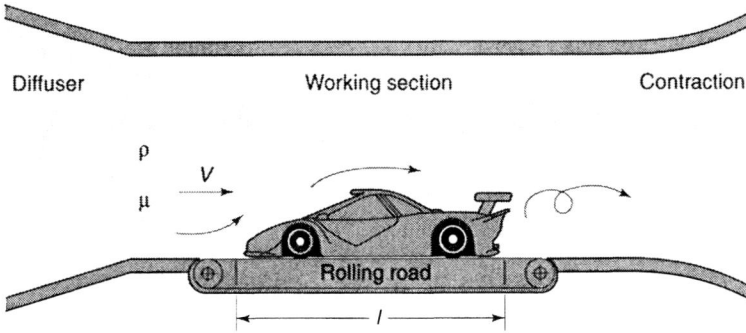

Figure E3.10

SOLUTION

We would normally start by carrying out the basic dimensional analysis of the problem. In this case we can use the result just obtained, i.e.

$$\frac{D}{\rho V^2 l^2} = F\left(\frac{\rho V l}{\mu}\right).$$

Dynamic similarity requires that each of the two non-dimensional groups has the same value for the model and the full-scale car, i.e.

$$C_D \equiv \frac{D_M}{\rho_M^2 V_M^2 l_M^2} = \frac{D_F}{\rho_F V_F^2 l_F^2}$$

and

$$Re \equiv \frac{\rho_M V_M l_M}{\mu_M} = \frac{\rho_F V_F l_F}{\mu_F}$$

where the subscripts M and F refer to the model and fullscale, respectively.

It is always advisable in problems of this sort to tabulate the known and unknown quantities:

	Model	*Full scale*
Speed V (m/s)	?	98.0 (356 kph)
Length l (m)	$l_F/4$	l_F
Air density ρ (kg/m^3)	4.7	1.2
Air viscosity μ (Pa.s)	1.7×10^{-5}	1.8×10^{-5}
Drag force D (N)	1334	?
Power P (W)	?	?

Note that we have no information about the absolute size of either the model or the full-scale car, only the ratio between them which, in this case, is sufficient.

From the Reynolds number equivalence we have

$$V_M = \frac{\rho_F}{\rho_M} \times \frac{\mu_M}{\mu_F} \times \frac{l_F}{l_M} \times V_F$$

$$= \frac{1.2}{4.7} \times \frac{1.7 \times 10^{-5}}{1.8 \times 10^{-5}} \times 4 \times 98.9$$

$$= 95.4 \ \text{m/s}.$$

From the drag coefficient we have

$$D_F = \frac{\rho_F}{\rho_M} \times \left(\frac{V_F}{V_M}\right)^2 \times \left(\frac{l_F}{l_M}\right)^2 \times D_M$$

$$= \frac{1.2}{4.7} \times \left(\frac{98.9}{95.4}\right)^2 \times 4^2 \times 1334$$

$$= 5859 \ \text{N or } 5.86 \ \text{kN}$$

and the corresponding tractive power required is

$$P_F = D_F V_F = 5859 \times 98.9$$

$$= 5.79 \times 10^5 \ \text{W}$$

$$= 579 \ \text{kW or } 777 \ \text{hp}.$$

Note that we did not consider scaling the tractive power directly and should ask the question, 'Does it also obey the dynamic similarity requirement?' A non-dimensional group including power is $P/\rho V^3 l^2$ (any other non-dimensional group including P would do as well) which for the model has the value

$$\frac{1334 \times 95.4 \times 16}{4.7 \times 95.4^3 \times l_F^2} = \frac{0.499}{l_F^2}.$$

For the full size we find

$$\frac{5.79 \times 10^5}{1.2 \times 98.9^3 \times l_F^2} = \frac{0.499}{l_F^2}$$

and the condition of dynamic similarity is again satisfied.

3.14 **Scaling complications**

The scaling situation we are usually confronted with is that of a model test on a scale much smaller than full size, and this can easily lead to conflicting or impossible requirements for the model tests. We illustrate the difficulty which can arise, and suggest ways in which the conflict can be resolved, by reconsidering the problem of aerodynamic drag. The observant reader will have noticed that in the sports-car case study (Example 3.10) the wind-tunnel air density was taken as 4.7 kg/m^3 and the dynamic viscosity as 1.7×10^{-5} Pa.s. These property values, which correspond to a pressure of about 3.7 bar and a temperature of 0°C, would be attainable only in a specially designed and very expensive wind tunnel. In some circumstances, strict adherence to the requirements of dynamic similarity is only possible through such extreme measures and it may be necessary to accept a compromise solution. Before going further, we should ask the question: 'What would be the consequences of performing the model test in a wind tunnel operating at normal temperature and pressure so that ρ_M and μ_M are the same as their full-scale counterparts, i.e. 1.2 kg/m^3 and 1.8×10^{-5} Pa.s, respectively?'

To answer the question just posed, we assume that the full-scale speed is still 356 kph, i.e. 98.9 m/s. For the model car the requirement of Reynolds number equality then leads to

$$V_M = \frac{\rho_F}{\rho_M} \times \frac{\mu_M}{\mu_F} \times \frac{l_F}{l_M} V_F$$

$$= 1 \times 1 \times 4 \times 98.9$$

$$= 395.6 \text{ m/s or } 1424 \text{ kph.}$$

It is immediately apparent that such an airspeed is unrealistically high and would again require a rather special (and again expensive) wind tunnel. There is, however, a more fundamental problem: since the speed of sound at 20°C is 343 m/s, a speed of 395.6 m/s corresponds to a Mach number $M = 1.15$. At this Mach number, changes in the airflow as it passed over the model would produce corresponding changes in pressure and density and compressibility effects, including shock waves, which would drastically affect aerodynamic behaviour. For the full-scale car, for which $M = 0.29$, compressibility effects would be (just about) negligible.

Clearly something has gone wrong and the foregoing is a reminder that when we carry out a dimensional analysis (or any other theoretical analysis) it is assumed that the physical quantities we have included account for all the physical effects of importance to the problem under consideration. Compressibility would be expected to influence the aerodynamic behaviour of rockets, missiles, shells, many aircraft, and cars such as Thrust 2 designed to challenge the land speed record, but not a car with a top speed of 'only' 356 kph. In the previous chapter we showed that both the isentropic bulk modulus $K \equiv \rho \, \partial p / \partial \rho \, |_s$ and the soundspeed c are suitable properties to characterise the compressibility of a fluid. Either property can be included in a dimensional analysis to account for compressibility but the soundspeed is usually the more convenient. In the aerodynamic drag example we can then write

$$D = f(V, l, \rho, \mu, c)$$

and the end result is

$$\frac{D}{\rho V^2 l^2} = F\left(\frac{\rho V l}{\mu}, \frac{V}{c}\right) \quad \text{or} \quad C_D = F(Re, M).$$

Note that we did not need to go through the entire dimensional analysis again: since we added one more variable, c, and no new dimensions, Buckingham's theorem tells us to look for one more non-dimensional group involving c. The conventional group is the Mach number $M = V/c$, though a Reynolds number using c rather than V (i.e. $\rho c l / \mu$) would be just as good.

Dynamic similarity now requires not only the Reynolds number Re to be the same for the model and the full-scale car, but also the Mach number M to be the same for both if the drag coefficient C_D is to be the same. As we saw in our example at normal temperature and pressure, Reynolds number equality was not consistent with Mach number equality and dynamic similarity was not achievable. We can see that this is always a potential problem if Reynolds number equality is enforced with the same fluid properties for both model and full scale because

$$\frac{\rho_M V_M l_M}{\mu_M} = \frac{\rho_F V_F l_F}{\mu_F}$$

leads to

$$V_M l_M = V_F l_F \quad \text{or} \quad V_M = \frac{l_F}{l_M} V_F$$

This result reveals as completely erroneous the common layman's assumption that the speed of a model car or aircraft is reduced in proportion to its size. In fact, the opposite is true!

The situation is much more satisfactory for the example with high pressure and reduced temperature for the model. The dynamic viscosity is practically independent of temperature and only slightly reduced by the pressure. Essentially the scale difference has been compensated for by increasing the density whereas the soundspeed (proportional to the square root of the absolute temperature for a perfect gas) and velocity are only slightly reduced and we find

$$M_M = M_F = 0.29.$$

The conditions for dynamic similarity are completely satisfied, although the Mach number is low enough for compressibility effects to be regarded as negligible. The more the full-scale Mach number exceeds 0.3 and approaches unity, the more important is equality of the model and full-scale Mach numbers to account for compressibility effects.

In general, the Reynolds number is affected by changes in both density and dynamic viscosity and it is often more convenient to define it in terms of the kinematic viscosity $v \equiv \mu/\rho$, i.e.

$$\frac{\rho V l}{\mu} = \frac{V l}{v}.$$

We can now regard the influence of the high pressure as a reduction in the kinematic viscosity. Such a reduction can also be achieved by changing the model fluid from air to a liquid such as water. The density of water may be taken as 1000 kg/m^3 and its dynamic viscosity at 20°C as 10^{-3} Pa.s so that its kinematic viscosity is $10^{-6} \text{ m}^2/\text{s}$, which is much lower than the value of $1.5 \times 10^{-5} \text{ m}^2/\text{s}$ for air at 1 bar, 20°C. For Reynolds number equality we now have

$$\frac{V_M l_M}{10^{-6}} = \frac{V_F l_F}{1.5 \times 10^{-5}} \quad \text{or} \quad V_M = \frac{1}{15} \frac{l_F}{l_M} V_F$$

i.e. if $l_F/l_M = 4$ we require $V_M = 0.27 V_F$ or $V_M = 26.7 \text{ m/s}$ in the sports-car example. Although a velocity of 26.7 m/s is much too high for a water channel and would probably introduce the new problem of cavitation (see section 8.10), it is clear that the low kinematic viscosity of water allows high Reynolds numbers to be achieved at relatively modest velocities.

We have already argued that if the Mach number is less than about 0.3, fluid compressibility is of negligible importance and Mach number equality is not vital. It is natural to ask if we can make a similar statement for the Reynolds number. If we calculate the Reynolds number for our sports car, taking the length as 5 m, we find $Re = 3.3 \times 10^7$, which is a very large number. In fact the Reynolds numbers for most situations of practical significance turn out to be quite large. It is also the case that above a critical Reynolds number, which is different for every body shape but typically of the order of 10^3, the flow becomes unsteady and increasingly random. Such a state of flow is given the name *turbulent* in contrast to that at very low Reynolds numbers where the fluid is smooth flowing and said to be *laminar*. In many instances, the drag coefficient in turbulent flow is almost constant, i.e. independent of Reynolds number, and so can be determined from model tests run at Reynolds numbers lower than full scale but still sufficiently high for the flow to be fully turbulent.

3.15 Summary

In this chapter we have explained the crucial role of units and dimensions in any problem involving physical quantities. The underlying principle of dimensional homogeneity has been introduced, i.e. the individual terms in any equation or function which connects physical quantities must have the same overall dimensions (and units). The major advantage of collecting the physical quantities, which arise in either a theoretical analysis or an experiment, into non-dimensional groups has been shown to be a reduction in the number of quantities which need to be considered separately. Buckingham's Π theorem was introduced as a method for determining the number of independent non-dimensional groups and the sequential elimination of dimensions as a systematic and simple procedure for identifying these groups. The scale up from a model to geometrically similar full-size equipment requires that the value of each of the non-dimensional groups describing the model-scale conditions is equal to its full-size counterpart to satisfy the condition of dynamic similarity. The definitions and names of the non-dimensional numbers most frequently encountered in fluid mechanics have been introduced and their physical significance explained. We concluded the chapter by pointing out that the conditions for dynamic similarity may be too costly or technically difficult to achieve in practice and a compromise solution has to be accepted.

The student should be able to:

- write down the units and dimensions of any of the physical quantities listed in Table A.6
- convert any physical equation into a non-dimensional form
- apply Buckingham's Π theorem to determine the number of non-dimensional groups
- use a systematic procedure, such as the method of sequential elimination of dimensions, to convert a functional dependence into a non-dimensional form
- recognise the more common non-dimensional groups which arise in fluid mechanics, such as Reynolds number, Mach number and drag coefficient
- determine full-size quantities from the results of a model-scale test and vice versa
- recognise and resolve scaling contradictions.

3.16 **Self-assessment problems**

3.1   Determine the dimensions of the following combinations of physical quantities:

$$\frac{p}{\rho V} \quad \frac{gt}{V} \quad \sqrt{\frac{g}{l}} \quad \frac{Vl}{v} \quad \rho V^2 \quad \frac{p}{gh} \quad \rho Vl$$

The symbols have the following meanings: p pressure; ρ density; V velocity; g acceleration due to gravity; t time; h, l lengths; v kinematic viscosity.

3.2   Find the values of the exponents, a, b, c which make each of the following combinations of physical quantities dimensionless

$$\frac{p}{\rho V^a} \quad \frac{\rho VD}{\mu^b} \quad \frac{D\sigma^c}{\rho v^2}.$$

The symbols have the following meanings: D diameter; p static pressure; V fluid velocity; ρ density; μ dynamic viscosity; v kinematic viscosity; σ surface tension.

3.3   A disk of weight W and radius R slides with velocity V over a flat surface. Lubricant of dynamic viscosity μ is pumped into the gap between the disk and the surface at a volumetric flow rate $\dot{Q}$. It can be shown that the drag force D acting on the disk is given by

$$D = \pi\left(\frac{\mu^2 R^4 W}{3\dot{Q}}\right)^{1/3} V.$$

Show that this equation is dimensionally correct.

3.4   A spherical drop of liquid of diameter D and density ρ oscillates under the influence of its surface tension σ. Show that the frequency of oscillation is given by

$$f = k\left(\frac{\sigma}{\rho D^3}\right)^{1/2}$$

where k is a constant.

3.5   (a) The wave resistance R of a ship depends upon its length L, the water depth d, the water density ρ, the acceleration due to gravity g and the ship speed V. Derive a non-dimensional form for the preceding sentence.
(b) A test carried out on a one-twentieth model ship of length 1.5 m in a towing tank gave a value for the wave resistance of 160 N for a model velocity of 4 m/s. The water density in the towing tank was 1000 kg/m³

while that for the full-size ship was 1050 kg/m^3. Calculate the depth of water in the towing tank to correspond to a depth of 20 m for the full-scale ship and the wave resistance for conditions of dynamic similarity. (Answers: 1 m, 1.34 MN)

3.6 The power P developed at the shaft of a windmill is a function of the wind speed V, the air density ρ, the rotation speed N and the windmill diameter D. Show that:

$$\frac{P}{\rho N^3 D^5} = \text{F}\left(\frac{V}{ND}\right).$$

Tests were performed on a model windmill and, as a result, values of $P/\rho N^3 D^5$ and the corresponding values of V/ND were obtained. It is required to design a windmill for a specified power, air density, wind speed and speed of the windmill. How would you plot and use the experimental results for this purpose?

3.7 (a) The shaft power P developed by a hydraulic turbine depends on the volume flow rate of liquid through the turbine $\dot{Q}$, the liquid density ρ, the pressure difference across the turbine Δp, the rotation speed N, and the rotor diameter D. Derive an appropriate non-dimensional form of this statement, ensuring that P, $\dot{Q}$ and Δp appear in independent non-dimensional groups.

(b) The volume flowrate of water through a turbine is $0.5 \text{ m}^3/\text{s}$ and the head difference is 40 m. If the turbine rotation speed is 250 rpm, calculate the flowrate and head difference for a geometrically similar quarter-scale model turbine running at 1200 rpm with a fluid of relative density 0.8 for conditions of dynamic similarity. Calculate also the ratio of power outputs for the two machines.

(Answers: $0.0375 \text{ m}^3/\text{s}$, 57.6 m, 11.6)

3.8 A rotating-paddle mixer is to be designed for use with a liquid of relative density 2 and a kinematic viscosity of $1 \text{ m}^2/\text{s}$. Laboratory tests show that the power required is 500 W for a model mixer with a paddle diameter of 100 mm operating at a rotational speed of 50 rpm in a fluid of density 1200 kg/m^3 and dynamic viscosity 100 Pa.s. Assuming dynamic and geometric similarity between the model and full-scale mixers, calculate the power and speed for a full-scale mixer with a paddle diameter of 1 m. (Answers: 144 kW, 6 rpm)

3.9 (a) The velocity V with which a shell can be fired from a gun barrel depends upon the shell diameter D, the shell mass m, the air density ρ, the speed of sound of the air c and the explosive energy of the shell E. Preferably using the method of sequential elimination of dimensions, derive a non-dimensional version of this statement.

(b) In a laboratory test, a shell-like projectile of diameter 10 mm and mass 0.05 kg is fired at a speed of 500 m/s through a heavy gas with a

density of 2 kg/m^3 and a sound speed of 100 m/s. The explosive energy required is 10 kJ. Calculate the shell speed, shell mass and explosive energy for a geometrically similar shell of 100 mm diameter fired into air with a density of 1.2 kg/m^3 and sound speed 340 m/s assuming conditions of dynamic similarity.
(Answers: 1700 m/s, 30 kg, 69.4 MJ)

3.10 (a) The drag force D on a supersonic aircraft may be assumed to depend on its wingspan S, its speed V, the air density ρ, and the compressibility of the air K. Preferably using the method of sequential elimination of dimensions, derive a non-dimensional form for this statement. Note that compressibility has the same dimensions as pressure.
(b) A particular aircraft is designed to fly at a speed of 2500 km/hr at an altitude where the air density is 0.287 kg/m^3 and the compressibility is 2.5×10^4 N/m^2. A 1/25th full-size model of the aircraft is to be tested in a pressurised wind tunnel. If the air in the tunnel has a density of 0.4 kg/m^3 and compressibility 5×10^5 N/m^2, calculate the air speed for dynamic similarity. If the model has a wingspan of 0.52 m and the drag force on the model is 6000 N, calculate the drag force on the full-size aircraft at design conditions.
(Answers: 2631 m/s, 187.5 kN)

3.11 (a) The drag force D exerted on a sphere moving through a fluid depends upon the sphere's radius R, the speed V of the sphere relative to the fluid, the fluid density ρ and the dynamic viscosity of the fluid μ. Use the method of sequential elimination of dimensions to derive a non-dimensional form of the previous sentence.
(b) Measurements of the drag force on six spheres moving through six different fluids yield the following results

	R (mm)	V (m/s)	ρ (kg/m^3)	μ (Pa.s)	D (N)
1	5	0.1	1260	1.3	0.0123
2	30	0.01	920	0.059	3.34×10^{-4}
3	2.5	0.5	935	0.12	2.83×10^{-3}
4	1	1	1000	10^{-3}	6.3×10^{-4}
5	7.5	10	1.2	1.8×10^{-5}	4.25×10^{-3}
6	50	120	0.07	8×10^{-6}	1.588

Use the results of part (a) to convert these results to non-dimensional form. Plot the logarithm of one of the non-dimensional groups against the logarithm of the other and comment on the results.

CHAPTER 4

Pressure variation in a fluid at rest

This chapter and the next are concerned with hydrostatics: the study of fluids at rest. The term *hydrostatics* is derived from the Greek word *hudor* meaning water but applies to all fluids. Shear stresses cannot arise in a body of fluid at rest because there is no relative motion (except at the molecular level) between fluid particles and the only internal forces that can occur are due to increases in pressure with depth which result from gravitational pull. In this chapter we shall derive the mathematical statement of this principle, the hydrostatic equation. The physics of many practical engineering problems involves nothing more than the application of the hydrostatic equation and in the second part of this chapter we shall apply it first to the measurement of pressure using liquid-filled tubes (manometry) and then to an analysis of pressure variations in the earth's atmosphere and in very deep water. We shall conclude the chapter by extending consideration to a body of fluid which, in spite of the chapter heading, is not in fact at rest but is in steady motion or even being accelerated but where there is still no relative motion between fluid particles.

4.1 Pressure at a point: Pascal's law

An important underlying principle of hydrostatics, first proved by the French scientist Blaise Pascal (1623–62), is that the pressure at any point in a body of fluid at rest is the same in all directions.

The proof of Pascal's law involves consideration of an infinitesimal (or elemental)* wedge of fluid, as shown in Figure 4.1. The wedge has thickness l, a vertical face of height δz, a horizontal face of length δx, and a sloping face of length δs. The fluid density within the wedge is ρ and the corresponding weight of the wedge is δW. At the outset we assume that the pressures which act on the

---

* The implication of the words 'infinitesimal' and 'elemental' here is that the wedge is so small that the pressure acting on any face of the wedge can be assumed to be uniform across that particular face, but may well vary from face to face. The concept of an infinitesimal element is commonly employed in fluid mechanics, and heat transfer. There is a lower limit on the size of such an element if the continuum hypothesis (section 2.3) is to remain valid. However, as we showed in section 2.3, in practice we rarely come close to this limit.

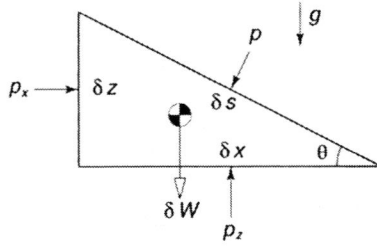

Figure 4.1 *Equilibrium of an infinitesimal wedge of fluid*

three faces of the wedge, p_x, p_z, and p, are all different. Since the fluid is at rest, the net force acting on the wedge in any direction must be zero.

For the horizontal direction we have

$$p_x \, \delta z \, l - p \, \delta s \, l \sin \theta = 0$$

and from the geometry of the wedge

$$\delta z = \delta s \sin \theta$$

so that

$$p_x \, \delta z \, l - p \, \delta z \, l = 0$$

or, after cancelling out $l \, \delta z$,

$$p_x = p. \qquad\qquad (4.1)$$

For the vertical direction we have

$$p_z \, \delta x \, l - p \, \delta s \, l \cos \theta - \delta W = 0$$

and, again from the geometry of the wedge,

$$\delta x = \delta s \cos \theta.$$

Since the volume of the wedge $\delta \vartheta$ is given by

$$\delta \vartheta = \tfrac{1}{2} \delta z \, \delta x \, l$$

its weight δW is

$$\delta W = \rho \, \delta \vartheta \, g = \tfrac{1}{2} \rho \, \delta z \, \delta x \, lg$$

so that

$$p_z \, \delta x \, l - p \, \delta x \, l - \tfrac{1}{2} \rho \, \delta z \, \delta x \, lg = 0$$

or

$$p_z = p + \tfrac{1}{2} \rho \, \delta z \, g - 0. \tag{4.2}$$

Equation (4.2) must hold no matter how small the wedge so that, as δz is reduced to zero, the term including δz must become negligibly small compared with p and p_z and we have

$$p_z = p. \tag{4.3}$$

Taken together, equations (4.1) and (4.3) allow us to conclude

$$p_x = p_z = p$$

i.e. since we have shrunk the wedge to nothing, *the pressure at any point in a fluid at rest has been shown to be the same in all directions.* This is Pascal's law.

4.2 Pressure variation in a fluid at rest: the hydrostatic equation

For a fluid at rest, the pressure is constant over any horizontal surface but increases with depth (or decreases with altitude, if you prefer). These statements of everyday experience are easily proved and quantified but have far-reaching implications.

Consider first a horizontal cylinder of fluid of infinitesimal length δx and infinitesimal cross-sectional area δA, as shown in Figure 4.2 (You might like to

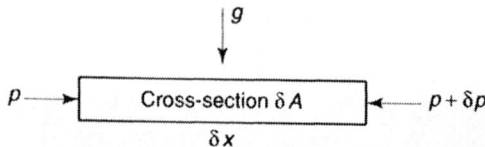

Figure 4.2 *Infinitesimal horizontal fluid cylinder*

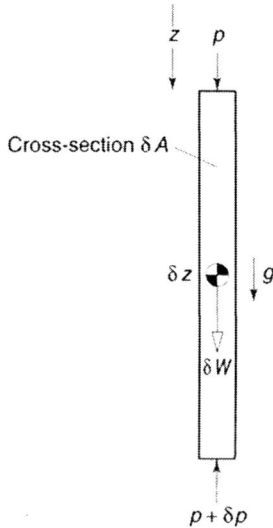

Figure 4.3 *Infinitesimal vertical fluid cylinder*

consider why the cross-section shape is unimportant.) If the pressure is assumed to change from p at one end of the cylinder to $p + \delta p$ at the other, as shown, then the net horizontal force acting on the fluid cylinder is

$$p\,\delta A - (p + \delta p)\,\delta A = -\delta p\,\delta A.$$

This net force must be zero unless the fluid cylinder is accelerating to the right ($\delta p < 0$) or to the left ($\delta p > 0$). Thus *for a fluid at rest* δp must be zero and we conclude that *pressure is constant along any horizontal line and, hence, over any horizontal surface.*

We consider now the forces acting on a vertical cylinder of fluid of infinitesimal length (i.e. height) δz and cross-sectional area δA, as shown in Figure 4.3. The key difference compared with the horizontal cylinder is that the force balance must now include the weight of the fluid cylinder δW acting vertically downwards.

The net downwards force acting on the fluid cylinder is

$$p\,\delta A + \delta W - (p + \delta p)\,\delta A = 0$$

so that, after cancellation of the term $p\,\delta A$ and rearrangement, we have

$$\delta p = \frac{\delta W}{\delta A}$$

i.e. the pressure increases by the amount δp due to the weight of fluid per unit area in a layer of fluid of depth δz. The cylinder weight is given by

$$\delta W = \rho \; \delta\vartheta \; g = \rho \; \delta z \; \delta A \; g$$

so that

$$\delta p = \rho \; \delta z \; g \quad \text{or} \quad \frac{\delta p}{\delta z} = \rho g.$$

If we now reduce δz towards zero, the finite difference ratio $\delta p / \delta z$ must by definition approach the pressure gradient dp/dz and we have the first-order ordinary differential equation

$$\boxed{\frac{dp}{dz} = \rho g.} \tag{4.4}$$

Equation (4.4) is known as the *hydrostatic equation*. Since both the fluid density ρ and the gravitational acceleration g are always positive, we conclude from equation (4.4) that dp/dz is always positive so that *the pressure in a body of fluid at rest can only increase with depth z (or decrease with altitude, $-z^*$).*

4.3 **Constant-density fluid**

It is almost always sufficient to regard the gravitational acceleration g as a constant, the value 9.81 m/s^2 normally being adopted for medium latitudes. In many engineering calculations sufficient accuracy is achieved if the fluid density ρ is also assumed to be constant. In these circumstances, equation (4.4) can be integrated to give

$$p = \rho g z + C \tag{4.5}$$

where C is a constant of integration which has to be determined from a boundary condition which provides a value for p at a known depth. It is frequently convenient to take the pressure at the origin for z (i.e. $z = 0$) as the barometric pressure B so that

$$\boxed{p = B + \rho g z.} \tag{4.6}$$

---

* Note that we have deliberately chosen z to be positive downwards to ensure that p increases with z. Had we chosen to adopt the sign convention that z is positive upwards, it would only change the mathematics, not the physics: for a fluid at rest, p must always increase with depth.

We shall refer to the combination $\rho g z$, which appears frequently throughout this book, as the *hydrostatic pressure*, and $p - \rho g z$ as the *piezometric pressure*.

As we discussed in Chapter 2, gases are far more compressible than liquids. That is to say, if the pressure to which a fixed mass of gas is subjected increases then, in general, its volume decreases and its density increases. For a gas at rest any pressure change can only result from a difference in altitude. As a consequence, even for a gas the constant-density approximation is frequently of sufficient accuracy if the height difference over which equation (4.5) or (4.6) is applied is limited to about 100 m. For liquids these equations are perfectly adequate for depths well in excess of 1000 m (see section 4.13). We may also note that equation (4.5) could also be applied to a solid, although p would then normally be referred to as a compressive normal stress. For a gas, unlike a solid, the pressure is always a compressive normal stress and can never become negative (i.e. tensile): the lowest possible absolute pressure is vacuum ($p = 0$). Since some gauges measure the pressure with respect to atmospheric, i.e. $p - B$, the so-called *gauge pressure* p_G, confusion sometimes arises if the pressure is sub-atmospheric (i.e. $p < B$) because the gauge pressure is then negative.

EXAMPLE 4.1
As shown in Figure E4.1, the cistern for a domestic central heating system is located in a loft 8 m above the boiler. Calculate the pressure of the water in the boiler and also the reading (gauge pressure) of the gauge attached to the boiler. Take the atmospheric pressure to be 1.02 bar.

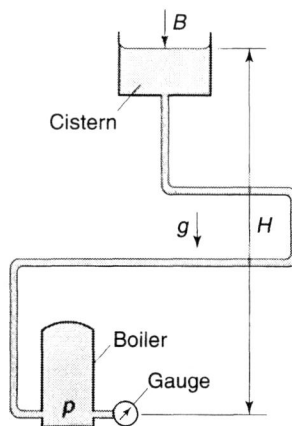

Figure E4.1

SOLUTION

$H = 8$ m; $\quad B = 1.02 \times 10^5$ Pa; $\quad g = 9.81$ m/s^2; $\quad \rho = 10^3$ kg/m^3

The boiler pressure p is given by

$$p = B + \rho g H$$
$$= 1.02 \times 10^5 + 10^3 \times 9.81 \times 8$$
$$= 1.80 \times 10^5 \text{ Pa} \quad \text{or} \quad 1.80 \text{ bar.}$$

Gauge pressure

$$p_G = p - B = 1.80 - 1.02$$
$$= 0.78 \text{ bar.}$$

Comment: Due to bends in the pipework connecting the cistern to the boiler, the length of piping may be much greater than the vertical height H. *Under static conditions, only the vertical height is important in determining the pressure difference $p - B$.*

4.4 Basic pressure measurement

In section 4.3 we found that for a fluid of constant density ρ the pressure increases linearly with vertical depth z according to

$$p = \rho g z + C \tag{4.5}$$

where C is a constant. If we apply equation (4.5) to a vertical tube containing a column of liquid of density ρ_M and height h, as shown in Figure 4.4, with the

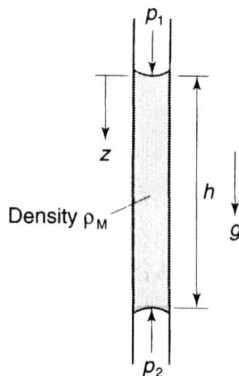

Figure 4.4 *A vertical tube containing a liquid column*

origin for z taken as the upper meniscus, we have

$$p_1 = C \qquad p_2 = \rho_M gh + C$$

or

$$\boxed{p_2 - p_1 = \rho_M gh.} \tag{4.7}$$

This is an important equation because it provides the basis for *manometry* which is *the measurement of pressure or pressure difference using manometers* (i.e. liquid-filled tubes).

4.5 Mercury barometer

If the upper surface of the liquid in the tube shown in Figure 4.4 is exposed to vacuum, the pressure p_1 is zero and we have a device called a barometer for measuring the absolute pressure p_2. Unfortunately life is rarely straightforward and this principle has a major flaw: as we discussed in section 2.12, at room temperature liquids tend to boil at pressures which are very low but still well above absolute zero making them unsuitable for use as a barometer liquid. In the case of water, for example, the pressure at which boiling begins, called the saturated vapour pressure p_V, is of course 1.013 bar at 100°C but still as high as 23 mbar at 20°C. The universal use of mercury as the barometer liquid is due to the Italian scientist Evangelista Torricelli (1608–47) who must have realised that its vapour pressure is negligibly low for all practical purposes, though still not identically zero (1.1×10^{-3} Pa or about 10^{-8} bar).

The basic arrangement of the classical mercury barometer is shown in Figure 4.5. The free surface of the mercury pool is exposed to the prevailing

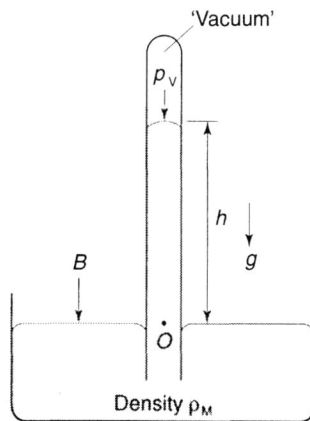

Figure 4.5 *Basic mercury barometer*

atmospheric pressure B so that

$$B = p_V + \rho_M gh. \tag{4.8}$$

Although we have just pointed out that it is negligible, for completeness we have included p_V, the saturated vapour pressure for mercury, in equation (4.8). Note too that we have made use of the fact that the pressure at point O within the barometer tube, on the same horizontal level as the free surface of the mercury pool, must be exactly the same as the atmospheric pressure. The density of mercury is usually taken as 13.6 kg/m^3, so that for the standard atmosphere, for which $B = 1.013$ bar, equation (4.8) gives $h = 0.76$ m or 760 mm or 29.9 in of mercury. *Any pressure can be converted into an equivalent height of any liquid* (ignoring the impracticality associated with vaporisation at low pressures) and, even in this age of metrification, it is still not uncommon to find pressures quoted in feet of water (1 bar is equivalent to 33.4 ft of water). The principal application of the barometer is to measure the small deviations from 1.013 bar which are such an important guide to forthcoming changes in the weather.

4.6 **Piezometer**

The piezometer is the second direct application of a column of liquid in a vertical tube for the measurement of pressure. In this case, as shown in Figure 4.6, the upper end of the tube is open to atmosphere (or any reference pressure) while the

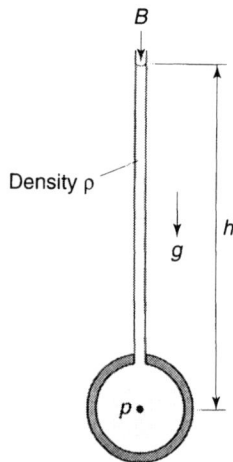

Figure 4.6 *Piezometer tube*

lower end is attached to a vessel or pipe containing a pressurised (i.e. $p > B$) liquid. The height of the liquid column h is related to the external atmospheric pressure B and the liquid pressure p according to the relation

$$p = B + \rho g h.$$

In fact what is measured here is effectively $\rho g h = p - B = p_G$, i.e. the gauge pressure. Note that the liquid depth within the vessel or pipe may represent a significant contribution to the overall height h.

4.7 U-tube manometer

As is probably apparent to the reader, the piezometer is not a very practical device for pressure measurement: there is the ever-present danger of liquid being blown into the environment if the tube is too short, but to cope with high pressures the vertical tube becomes excessively long, e.g. for water at a pressure of 10 bar h is 91.7 m. One solution to these problems might appear to be to place a high-density liquid, such as mercury, in the tube on top of the original liquid. In the case of water at 10 bar, for example, the corresponding height of a mercury column is 'only' 6.7 m. Unfortunately this idea is also impractical for a more fundamental reason: a heavy fluid on top of a light fluid is unstable. The mercury would simply run down into the main pipe or pressure vessel and create a major clean-up problem.

The U-tube arrangement shown in Figure 4.7 uses the heavy liquid (density ρ_M) idea but avoids the stability problem since the lighter liquid is above the

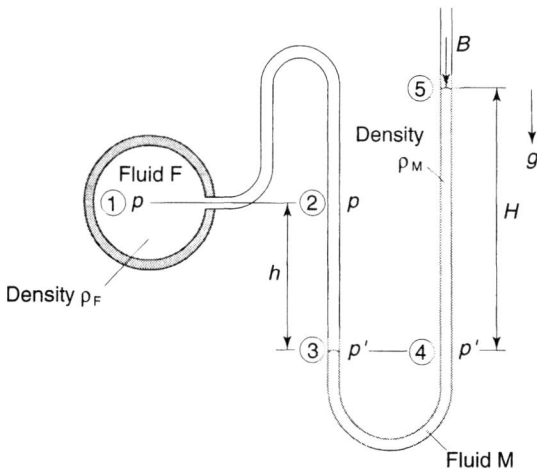

Figure 4.7 *U-tube manometer*

heavier liquid. The analysis of this and any other manometer problem depends upon the two fundamental results we obtained in sections 4.2 and 4.4:

(a) In a fluid at rest the pressure is the same for all points on the same horizontal level.
(b) For a fluid of constant density ρ_M, the pressure increase due to a *vertical* height difference h is $\rho_M gh$.

The first of these two statements tells us that in the fluid of density ρ_F the pressure at points ① and ② must be the same, and in the manometer liquid of density ρ_M the pressure at points ③ and ④ must be the same. Although we have no real interest in the intermediate pressure at the interface between the two fluids (point ③), it is often convenient in any analysis of this type to give it a symbol, such as p'.

For the right-hand side of the manometer we have

$$p' = B + \rho_M gH$$

and for the left-hand side

$$p' = p + \rho_F gh.$$

Note that in both cases we have evaluated the intermediate pressure p' by working our way down the manometer (i.e. in the direction of increasing pressure) by adding together the appropriate pressure and pressure difference for each fluid. This is a convenient and foolproof 'bookkeeping' approach to manometer problems which is easily extendable to any number of fluid layers. Since the right-hand side of both of the above expressions is equal to p', we can eliminate the intermediate pressure p' and write

$$B + \rho_M gH = p + \rho_F gh \quad \text{or} \quad p - B = (\rho_M H - \rho_F h)g$$

which gives us the unknown pressure p, once again in the form of the gauge pressure $p - B$. In fact, the pressure imposed on the free surface of the manometer liquid (point ⑤) need not be the barometric pressure B but could be any reference (i.e. known) pressure.

EXAMPLE 4.2
The U-tube manometer shown in Figure E4.2 is used to measure the pressure difference Δp between two pipes on the same horizontal level. Show that

$$\Delta p = (\rho_M - \rho_F)gH$$

where ρ_M is the density of the manometer liquid, ρ_F is the density of the fluid in the pipes, g is the acceleration due to gravity and H is the height

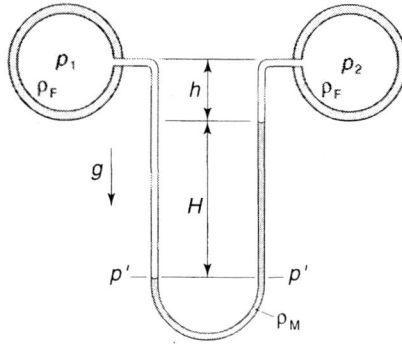

Figure E4.2

difference between the levels of the manometer fluid in the two arms of the U-tube.

SOLUTION
In the figure the fluid pressures in the two pipes have been indicated by p_1 and p_2, where $p_1 = p_2 + \Delta p$, the height of the pipes above the manometer fluid interface on the right-hand side as h, and the pressure in the manometer liquid on the horizontal level of the left-hand interface as p'.

For the left-hand side of the manometer we have

$$p' = p_1 + \rho_F g(h + H)$$

and for the right-hand side

$$p' = p_2 + \rho_F gh + \rho_M gH$$

If we equate these two expressions for p' we have

$$p_1 + \rho_F g(h + H) = p_2 + \rho_F gh + \rho_M gH$$

i.e.

$$p_2 + \Delta p + \rho_F gh + \rho_F gH = p_2 + \rho_F gh + \rho_M gH$$

since

$$p_1 = p_2 + \Delta p.$$

The term $p_2 + \rho_F gh$ cancels out and we have

$$\Delta p = (\rho_M - \rho_F)gH.$$

Comment: Neither the actual fluid pressures nor the height of the pipes appear in the final result which can be derived directly from the final equation of section 4.7 by setting $p - B = \Delta p$ and $h = H$.

4.8 Effect of surface tension

Until now we have neglected the effect of surface tension (see section 2.13) though, for liquids in tubes of small diameter, this property leads to an additional pressure difference which needs to be accounted for. To illustrate the point we consider water and mercury.

For pure water in contact with air the surface tension at 20°C is 7.28×10^{-2} N/m with a contact angle (measured through the liquid) which is practically zero. For mercury the surface tension is 4.72×10^{-1} N/m and the contact angle is 130°. What this means is that for water in a vertical tube, surface tension produces a force pulling vertically upwards whereas for mercury the resultant force is vertically downwards (due to symmetry there can be no net radial component of force). These two situations are illustrated in Figure 4.8.

For a liquid with surface tension σ and contact angle θ in a tube of diameter D, the magnitude of the resultant surface-tension force F is given by

$$F = \pi D \sigma \cos \theta$$

and for static equilibrium this force must exactly balance the weight of the liquid column of height h, i.e.

$$\pi D \sigma \cos \theta = \rho \tfrac{1}{4} \pi D^2 h g$$

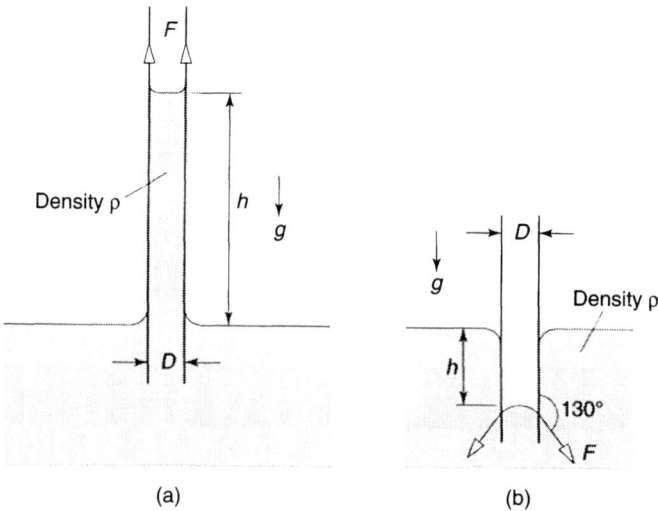

Figure 4.8 *Effect of surface tension: (a) water (b) mercury*

so that

$$h = \frac{4\sigma \cos \theta}{\rho g D}.$$

For water in a tube of diameter 1 mm we find $h = 29.7$ mm, which is clearly far from negligible, and even for a 10 mm tube we still have $h = 3$ mm. For mercury the corresponding results are -9.1 and -0.9 mm, respectively. Although the surface tension for mercury is about seven times that for water, its effect on the liquid level is much smaller, first because the density of mercury is 13.6 times that of water and second because the vertical component of the surface-tension force is reduced by the contact angle ($\cos 130° = -0.643$). The other obvious difference between the two liquids, due to their differing contact angles, is that water rises in a small-bore (or capillary) tube while the surface of mercury is pulled (not pushed) down.

Although it is straightforward to account for the effect of surface tension on the liquid in a manometer, it is usually unnecessary because we almost always measure changes in the liquid levels produced by a change in pressure difference so that the surface-tension effect cancels out.

4.9 Inclined-tube manometer

A common form of manometer for laboratory use is shown in Figure 4.9. Although a pressure difference $p - p_{REF}$ applied to the manometer produces changes in the vertical levels of the manometer liquids, in the inclined tube the level change H is determined by measuring the length L along the tube, i.e.

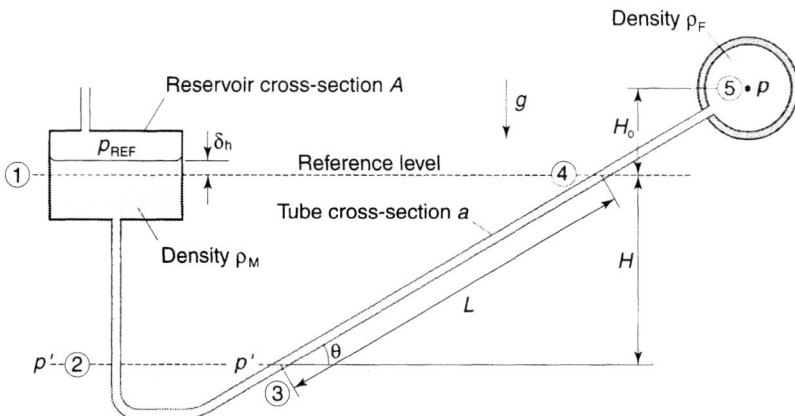

Figure 4.9 *Inclined-tube manometer*

$L = H/\sin\theta$ so that H is effectively amplified by the factor $1/\sin\theta$. Also, as we shall show shortly, the length L is further increased by minimising level changes on the reference (left-hand) side through the incorporation of a reservoir which has a cross-sectional area A much larger than that of the inclined tube a.

The analysis of the inclined-tube manometer is very similar to that of the U-tube manometer we considered in section 4.7. We measure all liquid levels with respect to a reference level, defined for convenience such that the manometer liquid in the reservoir is at the same level ① as that at ④ in the inclined tube before a pressure difference is applied. If the corresponding pressure in the pipe to which the inclined tube is connected is then p_0, we have

$$p_{REF} = p_0 + \rho_F g H_0 \tag{4.9}$$

where H_0 is the vertical height of the pipe centre (level ⑤) above the reference level. It is evident from the equation above that unless ρ_F is negligibly low, as would be the case for a gas at low pressure, p_0 may be appreciably different from the reference pressure in the reservoir p_{REF}.

For the situation shown in Figure 4.9, we again make use of the fact that the pressure in a single fluid at rest must be the same at all positions on the same horizontal level (i.e. in this case p' at level (②–③). On the left-hand side, again working vertically downward, we have

$$p' = p_{REF} + \rho_M g(\Delta h + H)$$

where Δh is the small change in height of the manometer liquid in the reservoir. On the right-hand side the corresponding result is

$$p' = p + \rho_F g(H_0 + H)$$

so that, after eliminating the intermediate pressure p' by subtraction, we have

$$p - p_{REF} = \rho_M g(\Delta h + H) - \rho_F g(H_0 + H).$$

This equation reveals the reason for incorporating into the manometer a reservoir of large cross-section: the smaller we can make Δh, the larger will be H and so L, and it is usually easier to measure accurately a long length than it is a short length. Since it would be inconvenient and inaccurate if we had to measure the small level change Δh, we make use of the fact that this height change corresponds to the movement of a liquid volume in the reservoir $A\,\Delta h$ which must exactly equal the volume of manometer liquid pushed down* the inclined tube

---

* Although we have effectively assumed that an increase in the pressure p acts to force the manometer liquid down the inclined tube, the analysis is equally valid for movement up the tube corresponding to a pressure decrease, although L is then negative.

aL. It is important to understand that whereas the vertical height difference is equal to $L \sin \theta$, the volume of liquid for the inclined tube is aL not $aL \sin \theta$.

Equality of the displaced volumes gives

$$A \, \Delta h = aL$$

and we also have

$$H = L \sin \theta$$

so that the equation for $p - p_{REF}$ becomes

$$p - p_{REF} = \rho_M g \left(L \sin \theta + \frac{La}{A} \right) - \rho_F g (L \sin \theta + H_0)$$

or

$$p - p_{REF} = (\rho_M - \rho_F) gL \sin \theta + \frac{\rho_M gLa}{A} - \rho_F g H_0.$$

It is usual for the manometer to be designed such that the cross-section of the inclined tube a is so much smaller than that of the reservoir A that the term $\rho_M gLa/A$ can be neglected compared with $(\rho_M - \rho_F) gL \sin \theta$. In fact the term involving a can easily be taken into account as a correction term but ignored at the time of making the measurement. As mentioned earlier, it may also be that the density of the manometer liquid ρ_M far exceeds the density of the fluid in the pipe ρ_F (e.g. for mercury and air at normal temperature and pressure the density ratio ρ_M/ρ_F is 1.13×10^4), and the manometer equation simplifies to

$$p - p_{REF} \approx \rho_M gL \sin \theta - \rho_F g H_0.$$

For the reason just given, the final term may also be negligible, or it may be that we are only interested in changes in pressure compared with p_0, in which case we can substitute for p_{REF} from equation (4.9) to obtain

$$p - p_0 \approx \rho_M gL \sin \theta.$$

Example 4.3

The inclined manometer shown in Figure E4.3 is used to measure the difference between the pressure p in a horizontal pipe and the constant reference pressure p_R in the manometer reservoir. The reference level is defined by the surface of the manometer liquid in the reservoir being at the same height as the interface (in the inclined tube) between the pipe and manometer liquids. The pipe axis is a vertical distance H above the reference

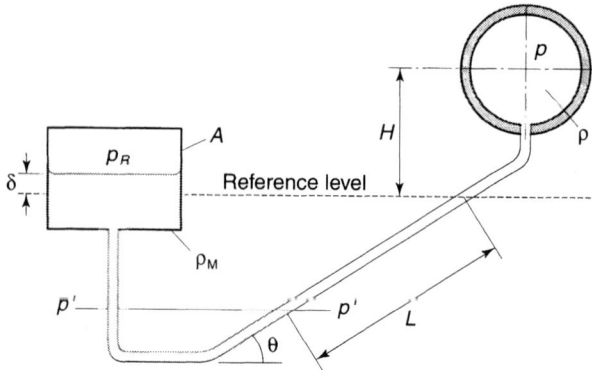

Figure E4.3

level. Derive a relationship between $p, p_R, H, g, L, \theta, \rho_M, \rho, a$ and A, where g is the acceleration due to gravity, the manometer reading L is the distance moved from the reference level by the interface in the inclined tube, ρ_M is the density of the manometer liquid, ρ is the density of the fluid in the pipe, a is the cross-sectional area of the inclined tube and θ its angle of inclination, and A is the cross-sectional area of the reservoir.

For a particular manometer, $p_R = 0.5$ bar, $H = 0.5$ m, $\theta = 20°$, $\rho_M = 13\ 600$ kg/m^3, $\rho = 800$ kg/m^3, the internal diameter of the inclined tube is 5 mm, and that of the reservoir 200 mm. Calculate the manometer reading L if the reservoir and pipe pressures are equal. Calculate the pressure difference $p - p_R$ if $L = 0$. Calculate the saturation vapour pressure of the liquid in the pipe if cavitation occurs in the pipe for $L = -0.9$ m.

SOLUTION

We note first that the pressure p' at the level of the interface is the same in both arms of the manometer.

On the left-hand side

$$p' = p_R + \rho_M g(\delta + L \sin \theta)$$

where δ is the vertical height of the manometer fluid in the reservoir above the reference level. Also, $L \sin \theta$ is the vertical change in height due to movement of the manometer liquid in the sloping tube.

On the right-hand side

$$p' = p + \rho g(H + L \sin \theta).$$

We can eliminate p' to give

$$p - p_R = \rho_M g (\delta + L \sin \theta) - \rho g (H + L \sin \theta)$$
$$= (\rho_M - \rho) g L \sin \theta + \rho_M g \delta - \rho g H.$$

Since the volume of manometer liquid which moves above the reference level on the left-hand side must equal the volume of liquid which moves down the inclined tube, we have

$$A\delta = aL$$

so that

$$p - p_R = (\rho_M - \rho) g L \sin \theta + \left(\frac{\rho_M a L}{A} - \rho H \right) g$$

which is the required result.

For the numerical part of the problem we have

$p_R = 5 \times 10^4$ Pa; $H = 0.5$ m; $\theta = 20°$; $\rho = 800$ kg/m³;

$\rho_M = 1.36 \times 10^4$ kg/m³; $d = 5 \times 10^{-3}$ m; $D = 0.2$ m; $g = 9.81$ m/s²

If the reservoir and pipe pressures are equal $p = p_R$, and we have

$$(\rho_M - \rho) L \sin \theta + \frac{\rho_M a L}{A} - \rho H = 0$$

or

$$L = \frac{\rho H}{(\rho_M - \rho) \sin \theta + (\rho_M a / A)}$$

$$= \frac{800 \times 0.5}{(1.36 \times 10^4 - 800) \sin 20° + 1.36 \times 10^4 \times (5/200)^2}$$

$$= 0.091 \text{m} \quad \text{or} \quad 91 \text{ mm}.$$

For $L = 0$, the pressure difference is given by

$$p - p_R = -\rho H g$$
$$= -800 \times 0.5 \times 9.81$$
$$= -3924 \text{ Pa}$$

For $L = -0.9$ m, the pressure difference is given by

$$p - p_R = -(1.36 \times 10^4 - 800)9.81 \times 0.9 \times \sin 20°$$

$$-\left[1.36 \times 10^4 \times 0.9 \times \left(\frac{5}{200}\right)^2 + 800 \times 0.5\right]9.81$$

$$= -4.265 \times 10^4 \text{ Pa}.$$

Thus

$$p = 5 \times 10^4 - 4.265 \times 10^4$$
$$= 7.35 \times 10^3 \text{ Pa}.$$

Since this is the pressure at which we are told cavitation occurs, we conclude that the vapour pressure $p_V = 7350$ Pa.

4.10 Multiple fluid layers: stratification

Fresh water has a lower density (normally taken as 1000 kg/m³) than salt water
($\approx 1000 + 7c$ kg/m³ where c is the salt concentration percentage by weight), so if
one meets the other, as in an estuary, the lighter fresh water tends to form a layer
above the heavier salt water. This phenomenon of layering according to progressively increasing density, which can occur in both liquids and gases, is called
stratification and can be of major practical significance. For example, under
certain atmospheric and topographical conditions, as frequently occur in the Los
Angeles basin, a 'lid' of low-density warm air settles above heavy cooler air and
leads to the build up of dangerous levels of pollutants at ground level.

If there are several layers of fluid, each of a different constant density, the
pressure increases across each layer simply add together to give the total increase
in pressure with depth. Using the notation of Figure 4.10, we have

$$p_1 = B + \rho_1 g Z_1$$
$$p_2 = B + \rho_1 g Z_1 + \rho_2 g Z_2$$
$$p_3 = B + \rho_1 g Z_1 + \rho_2 g Z_2 + \rho_3 g Z_3.$$

The linear increase in pressure with depth corresponding to $p = \rho g z + C$ is shown
schematically in each layer.

EXAMPLE 4.4

A vertical cylinder of inside diameter 50 mm is sealed at the bottom
and filled to a depth of 500 mm with mercury. If the barometric pressure is

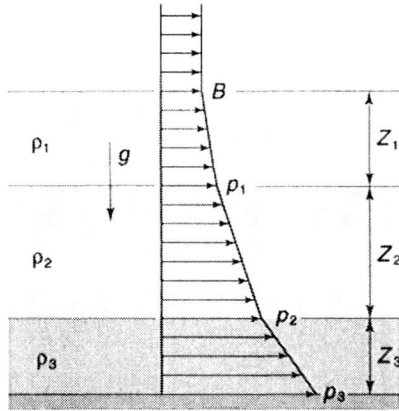

Figure 4.10 *Pressure increase through a stratified fluid*

1.1 bar and the mercury supports a piston of mass 5 kg, calculate the absolute and gauge pressures at the bottom of the cylinder.

SOLUTION

$D = 0.05$ m; $B = 1.1 \times 10^5$ Pa; $m = 5$ kg; $H = 0.5$ m; $\rho = 13.6 \times 10^3$ kg/m^3 (note that we have converted the values of all physical quantities to basic SI units).

In the absence of any other information, we assume that the piston is a perfect fit in the cylinder, with no leakage or friction. We denote the cross-sectional area of the cylinder ($= \frac{1}{4}\pi D^2$) by A and the absolute pressure at

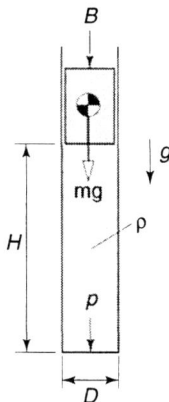

Figure E4.4

the bottom of the cylinder by p. The solution to this problem requires that we recognise that the force acting on the bottom of the cylinder pA is made up of three components:

$$pA = BA + mg + \rho gHA$$

from which

$$p = B + \frac{mg}{A} + \rho gH.$$

The effective pressure difference due to the piston Δp_P is obviously mg/A. If the density of the piston material is ρ_P and its height Z, then $m = \rho_P AZ$ and $\Delta p_P = \rho_P gZ$, i.e. a hydrostatic or 'ρgz' term just as for a fluid.

If we now substitute the numerical values into the equation for p, we have

$$p = 1.1 \times 10^5 + \frac{5 \times 9.81}{(\pi/4) \times 0.05^2} + 13.6 \times 10^3 \times 9.81 \times 0.5$$

$$= 2.02 \times 10^5 \text{ Pa} \quad \text{or} \quad 2.02 \text{ bar}$$

and the gauge pressure $p_G = p - B = 0.92$ bar.

4.11 Variable-density fluid

Since we put no restriction on the density ρ in the derivation of the hydrostatic equation (4.4) in section (4.1), this equation must apply not only as in the previous two sections, where ρ was taken either as constant throughout the body of fluid under consideration or constant within layers of fluid, but also where the density varies in some way with depth z, i.e. $\rho = \rho(z)$ which simply means 'ρ depends upon z'. It may be that the density variation with z is known and equation (4.1) can be integrated immediately. The more usual (and more complicated) situation is that the density is related to the fluid pressure (and possibly also the fluid temperature) and finding the density variation is itself part of the solution of the problem.

We consider first the simpler situation where the density is a specified function of depth. For example, we might assume that close to the bed of a reservoir increasing amounts of silt cause the effective fluid density ρ to increase linearly with depth, i.e.

$$\rho = \rho_0 + \alpha z$$

where ρ_0 is the density of pure water and α is a constant which depends upon the silt concentration. The hydrostatic equation is then

$$\frac{dp}{dz} = \rho g = (\rho_0 + \alpha z)g$$

which we can write as

$$dp = \rho_0 g \, dz + \alpha gz \, dz.$$

If we integrate between the levels z_1 and z_2, we have finally

$$p_2 - p_1 = \rho_0 g(z_2 - z_1) + \tfrac{1}{2} \alpha g(z_2^2 - z_1^2)$$

i.e. the pressure increase is made up of two parts, the first due to the pure water density ρ_0 and the second a quadratic (i.e. z^2) term due to the silt.

4.12 Earth's atmosphere

The earth's atmosphere is a relatively thin layer of gas held to the earth's surface by gravitational attraction. Although we rarely think of air as being heavy, the remarkable fact is that the total mass of the atmosphere is estimated to be 5.27×10^{18} kg, most of it contained within an 11 km thick layer called the *troposphere*, the first of a series of layers which make up the International Civil Aviation Organisation (ICAO) *Standard Atmosphere*. This Standard Atmosphere, which is an attempt to represent average atmospheric conditions in temperate latitudes, is assumed to consist of dry air with a sea-level (zero altitude) temperature of 15°C and pressure (the *standard atmospheric pressure*) of 1.013 bar. The properties of the Standard Atmosphere are listed in Table A.8.

In the troposphere the temperature decreases with *altitude z'* (in metres) according to the relation

$$T = T_0 - \kappa z'$$

where the *lapse rate* (i.e. temperature gradient) $\kappa = 6.5 \times 10^{-3}$ °C/m. The next layer, which extends from $z' = 11$ km (the 'top' of the troposphere) to 20.1 km, is the *stratosphere* in which the temperature remains constant at −56.5°C. For subsonic civil aircraft the typical cruising altitude is 10 km, whereas for supersonic and military aircraft it is more like 20 to 30 km. The low air temperature at high

---

* For convenience we use the symbol z' for altitude, with $z' = 0$ as sea level, rather than regard altitude as negative depth. The hydrostatic equation must now be written as $dp/dz' = -\rho g$, the negative sign appearing because $dz = -dz'$.

altitude is responsible for condensing the water vapour in the engine exhaust from aircraft to produce the white vapour trails called contrails often visible in a clear blue sky.

As stated in section 4.11, rather than knowing the density variation in advance, the more usual situation is that the density is related to the pressure and temperature of the fluid. Since it is the pressure we are trying to find, this situation is clearly more complicated: typically thermodynamic information is required, e.g. in the form of an equation of state, and the integration is slightly more difficult. This is the situation for the earth's atmosphere for which, as we have just seen, the basic specification is in terms of temperature, a quantity which is relatively easy to measure.

We illustrate the general problem by considering the stratosphere in which the temperature remains constant with altitude (a constant temperature process is termed *isothermal*). The connection between density, absolute pressure and absolute temperature (i.e. temperature in degrees kelvin) is the ideal-gas law, which was introduced in section 2.5:

$$p = \rho RT \qquad (4.10)$$

where R is the specific gas constant, usually taken as $287 \text{ m}^2/\text{s}^2.\text{K}$ for air.

Since the temperature T here is constant, it is convenient to use equation (4.10) to eliminate the variable density from the hydrostatic equation to give

$$\frac{dp}{dz'} = -\frac{pg}{RT} \quad \text{or} \quad \frac{1}{p} dp = -\frac{g}{RT} dz'$$

which can be integrated to give

$$\ln p = -\frac{gz'}{RT} + C$$

where C is a constant of integration which we can determine from the condition at the top of the troposphere ($z'_T = 11$ km) where the pressure $p_T = 0.226$ bar so that

$$C = \ln p_T + \frac{gz'_T}{RT}$$

and hence

$$\ln p = -\frac{gz'}{RT} + \ln p_T + \frac{gz'_T}{RT}$$

or

$$\frac{p}{p_T} = \exp\left[\frac{g(z'_T - z')}{RT}\right]. \qquad (4.11)$$

What equation (4.11) shows is that the air pressure in the stratosphere decreases exponentially with altitude difference. A similar analysis for the troposphere, for which the temperature decreases linearly with altitude, i.e. $dT/dz' = -\kappa$ where κ is the lapse rate, leads to the relation

$$p = B\left(1 - \frac{\kappa z'}{T_0}\right)^{g/\kappa R} \tag{4.12}$$

where T_0 is the temperature at sea level (taken as 288 K). Since $\kappa = 6.5 \times 10^{-3}$ °C/m, the exponent $g/\kappa R = 5.26$.

For altitudes beyond the stratosphere, starting with the *mesosphere*, the temperature initially increases at a rate of 1 K/km up to 32 km and at about 2 K/km up to 45 km. The properties of the standard atmosphere up to about 70 km are shown graphically in Figure 4.11 and tabulated in Table A.8. Both the pressure and the density decrease almost exponentially with altitude, each reaching nearly zero at $z' = 40$ km. If this figure is taken as the height of the atmosphere, an average density $\bar{\rho} = B/gz'$ can be calculated as 0.26 kg/m³, i.e. only about 20% of the sea-level density at 15°C and 1 atmosphere.

EXAMPLE 4.5

The temperature in the lower mesosphere ($20.1 < z' < 32.2$ km) increases linearly with altitude z'. If the gas which makes up the mesosphere can be treated as a perfect gas, show that the pressure varies according to

$$\frac{p}{p_S} = \left(\frac{T}{T_S}\right)^{-g/\kappa R} \tag{4.13}$$

where κ is the temperature gradient and the subscript S denotes the top of the stratosphere ($z'_S = 2.01$ km).

If the values of p_S, T_S and κ are 0.0548 bar, 216.5 K and 1°C/km, respectively, calculate the temperature, pressure and density at an altitude z' of 32.2 km. The gas constant R can be taken as 287 m²/s².K.

SOLUTION

The temperature variation with z' is linear, i.e.

$$\frac{dT}{dz'} = \kappa$$

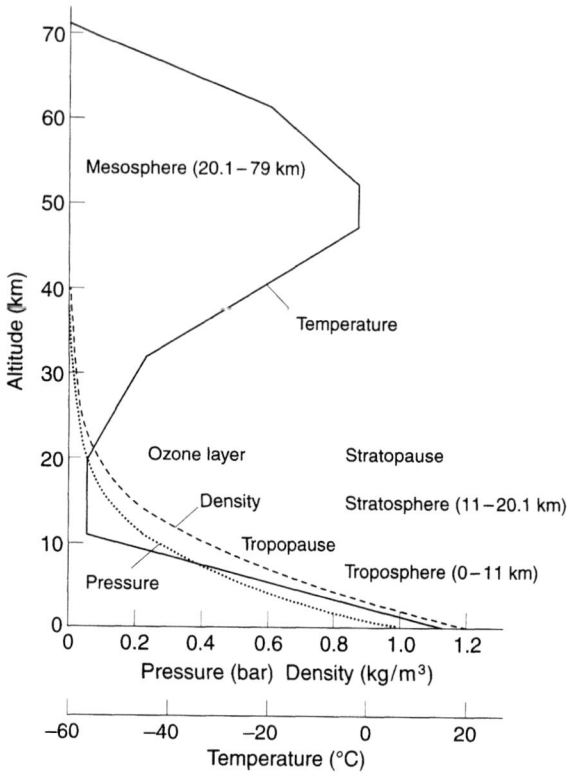

Figure 4.11 *Temperature, pressure and density variation for the ICAO standard atmosphere*

and the pressure follows the hydrostatic equation

$$\frac{dp}{dz'} = -\rho g.$$

These two equations can be combined to eliminate z' to give

$$\frac{dp}{dT} = -\frac{\rho g}{\kappa}.$$

The density ρ can be eliminated using the perfect-gas law

$$p = \rho RT$$

to give

$$\frac{dp}{dT} = -\frac{pg}{\kappa RT}.$$

which can be rearranged as

$$\frac{dp}{p} = -\frac{g}{\kappa R}\frac{dT}{T}.$$

This equation can be integrated to give

$$\ln p = -\frac{g}{\kappa R}\ln T + C.$$

The constant of integration C can be determined from the pressure and temperature at the top of the stratosphere, i.e.

$$\ln p_S = -\frac{g}{\kappa R}\ln T_S + C.$$

We can eliminate C by subtraction to give

$$\ln p - \ln p_S = -\frac{g}{\kappa R}(\ln T - \ln T_S)$$

or

$$\ln\left(\frac{p}{p_S}\right) = -\frac{g}{\kappa R}\ln\left(\frac{T}{T_S}\right),$$

so finally

$$\frac{p}{p_S} = \left(\frac{T}{T_S}\right)^{-g/\kappa R}.$$

For the numerical part of the problem we have $p_S = 5.48 \times 10^3$ Pa; $T_S = 216.5$ K; $R = 287$ m^2/s^2.K; $g = 9.81$ m/s^2; $z'_S = 20.1$ km; $\kappa = 10^{-3}$ °C/m; $z' = 3.22$ km

$$\frac{dT}{dz'} = \kappa$$

therefore

$$T = \kappa z' + C \quad \text{and} \quad T_S = \kappa z'_S + C$$

so

$$T - T_S = \kappa(z' - z_S').$$

Thus the temperature at $z' = 32.2$ km is

$$
\begin{aligned}
T &= 216.5 + 1.0(32.2 - 20.1) \\
&= 216.5 + 12.1 \\
&= 228.6 \ K \quad \text{or} \quad -44.4°C.
\end{aligned}
$$

The corresponding pressure is given by

$$\frac{p}{p_S} = \left(\frac{T}{T_S}\right)^{-g/\kappa R}$$

$$= \left(\frac{228.6}{216.5}\right)^{-9.81/(10^{-3} \times 287)}$$

$$= 0.156$$

so

$$
\begin{aligned}
p &= 0.156 \times 5.48 \times 10^3 \\
&= 854 \ Pa \quad \text{or} \quad 8.54 \ \text{mbar}
\end{aligned}
$$

and the density ρ is found from

$$
\begin{aligned}
\rho &= \frac{p}{RT} = \frac{854}{287 \times 228.6} \\
&= 0.0130 \ \text{kg/m}^3.
\end{aligned}
$$

4.13 Deep oceans

The deepest ocean has a depth of 10 550 m. The corresponding pressure at that depth, calculated from $B + \rho g h$ for a liquid of density 1000 kg/m³, would be 1036 bar. An ideal gas subjected to such an enormous pressure (at constant temperature) would decrease to 1/1036 of its volume at 1 bar with a corresponding increase in density. How realistic then is our assumption of constant density for water and other liquids? As stated in section 2.6, a good approximation for the observed behaviour of water is the relation

$$\frac{p + 3000}{3001} = \left(\frac{\rho}{1000}\right)^7 \qquad (4.14)$$

where the pressure p is measured in bar. For a pressure of 1036 bar this equation gives a density just 4.3% higher than that for 1 bar. We conclude that for water, at least, the increase in density with pressure is likely to be negligible in most practical circumstances. If necessary, however, we could combine equation (4.14) with the hydrostatic equation (4.4) to determine a more accurate $p(z)$ relation than that which results from the assumption of constant density.

4.14 Pressure variation in an accelerating fluid

Until now we have restricted consideration to the vertical variation of pressure in fluids at rest. In fact the crucial restriction which allows us to consider effects due only to pressure variations is that there should be no relative movement between the fluid particles and hence the shear stress is identically zero throughout the fluid. It should be self-evident that there is no relative movement within a fluid in a container moving at constant velocity, provided sufficient time has elapsed for any motion created at the start of the process to have died out. In these circumstances, everything we have said in this chapter so far still holds. As an example, every particle of petrol in a closed container in the boot of a car travelling at constant speed would be moving at the speed of the car. The situation of fluid subjected to constant acceleration is a little more complex.

As in section 4.2, we consider a horizontal cylinder of fluid of infinitesimal length δx and infinitesimal cross-section δA but now, as shown in Figure 4.12, with a constant component of acceleration a_x in the horizontal direction. If we apply Newton's second law of motion (i.e. net force = mass × acceleration) to the fluid cylinder in the x-direction, we have

$$p \, \delta A - (p + \delta p) \, \delta A = \rho \, \delta A \, \delta x \, a_x$$

wherein we have substituted $\delta A \, \delta x$ for the volume $\delta \vartheta$ of the elemental cylinder and $\rho \, \delta \vartheta$ for its mass.

This equation simplifies to

$$\frac{\delta p}{\delta x} = -\rho a_x$$

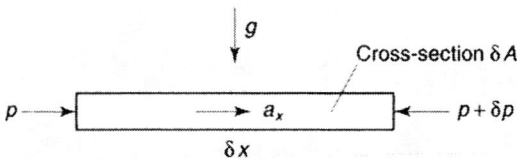

Figure 4.12 *Accelerating horizontal cylinder of fluid*

or, in the limit as δx approaches zero,

$$\frac{\partial p}{\partial x} = -\rho a_x. \tag{4.15}$$

The minus sign should come as no surprise: if pressure is higher to the right than to the left, we must expect that the fluid will accelerate to the left and not to the right as the acceleration vector in the figure would suggest. For the first time in this book we have used the symbols to denote a *partial derivative* $\partial/\partial x$ rather than a total derivative d/dx. The partial derivative is appropriate here because the hydrostatic pressure p can now vary both in the horizontal (i.e. x) direction as well as in the vertical (i.e. z) direction. If we also subject the fluid to a component of acceleration a_z vertically downwards, it is straightforward to show that the hydrostatic equation (4.4) must be replaced by

$$\frac{\partial p}{\partial z} = \rho(-a_z + g). \tag{4.16}$$

The spatial variation (i.e. with x and z) of the hydrostatic pressure requires the solution of both equations (4.15) and (4.16). Provided the acceleration components a_x and a_z are constant, both equations are easily integrated and we find

$$p = -\rho a_x x + C_1(z) \quad \text{and} \quad p = -\rho(a_z - g)z + C_2(x).$$

$C_1(z)$ and $C_2(x)$ are now functions of integration rather than the constants of integration we have for ordinary differential equations. Since these two equations for p must be simultaneously valid, the final result is

$$p - p_{REF} = \rho(gz - a_z z - a_x x) \tag{4.17}$$

where p_{REF} is a reference pressure (i.e. $p = p_{REF}$ at $x = 0$, $z = 0$).

If the terms within brackets in equation (4.17) are constant, we see that lines for which

$$gz - a_z z - a_x x = \text{constant}$$

represent lines of constant pressure (*isobars*). The slope of the isobars (see Figure 4.13) is given by

$$\tan \theta = \frac{dz}{dx} = \frac{OB}{OA} = \frac{a_x}{g - a_z}$$

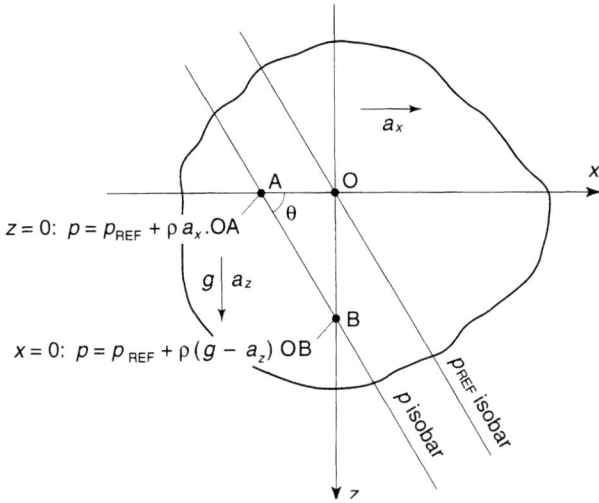

Figure 4.13 *Isobars in an accelerating body of fluid*

EXAMPLE 4.6

A Formula One racing car is driven around a corner of radius 64 m at a constant speed of 180 kph. What angle does the free surface of the petrol in the fuel tank make with the horizontal? Calculate the pressure difference between the free surface and a point a perpendicular distance 200 mm from the free surface. Take the density of petrol as 800 kg/m^3.

SOLUTION

Since the car speed $V = 180$ kph $= 50$ m/s, the centripetal (i.e. radially inward) acceleration for a radius of 64 m is

$$a_x = \frac{V^2}{R} = \frac{50^2}{64} = 39.1 \text{ m/s}^2 \qquad \text{(i.e. a lateral acceleration of almost } 4g\text{)}.$$

In this case the vertical acceleration a_z is zero, and so the slope of the free surface is given by

$$\tan \theta = \frac{a_x}{g} = 3.98$$

i.e.

$$\theta = 75.9°$$

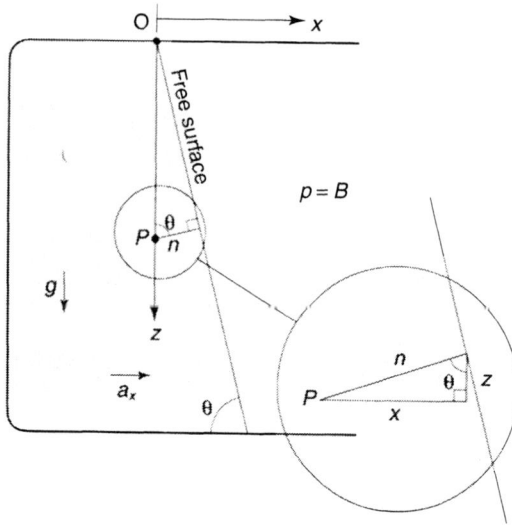

Figure E4.6

From Figure E4.6 we see that for the point P, 200 mm ($=n$) from the free surface,

$$x = -n \sin \theta \quad \text{and} \quad z = n \cos \theta$$

so that the pressure at P is given by

$$p - B = \rho g n \left(\frac{a_x}{g} \sin \theta + \cos \theta \right)$$

$$= 800 \times 9.81 \times 0.2(3.98 \times \sin 75.9° + \cos 75.9°)$$

$$= 6441 \text{ Pa} \quad \text{or} \quad 0.064 \text{ bar.}$$

4.15 SUMMARY

We started this chapter by establishing the three fundamental principles for the variation of pressure throughout a body of fluid at rest: (i) the pressure at a point is the same in all directions, (ii) the pressure is the same at all points on the same horizontal level and (iii) the pressure increases with depth z according to the hydrostatic equation $dp/dz = \rho g$. If the fluid density is constant, the increase in pressure over a depth increase h is $\rho g h$, a result which can be used to analyse the response of simple barometers and manometers to applied pressure changes. In situations where very large changes in pressure occur, such as throughout the earth's atmosphere and in very deep water, the assumption of constant density may be inadequate and an equation of state is then required to relate pressure and density together with an assumption about the fluid temperature. The hydrostatic equation is still valid but more difficult to integrate.

The student should be able to

- calculate the pressure variation with vertical depth for a fluid of constant density, including the situation of a series of fluid layers
- analyse the response of a simple barometer to changes in the external pressure
- analyse the response of a U-tube or inclined-tube manometer to changes in the applied pressure difference
- calculate the pressure variation with vertical depth (or height) for a variable-density fluid where there is a simple relationship between pressure and fluid density.

4.16 Self-assessment problems

4.1   If the atmosphere is assumed to have a height of 100 km and a constant density, calculate the density if the pressure at ground level is 1 bar. (Answer: $0.102 \, \text{kg/m}^3$)

4.2   An open tube of length 20 m sloping at an angle of 20° to the horizontal is half full of water and half full of oil of relative density 0.8. Calculate the hydrostatic pressure at the bottom of the tube. (Answer: 0.604 bar)

4.3   A vertical U-tube manometer contains two liquids of density ρ_0 and ρ, respectively. Show that the difference in height between the two liquids, when no pressure difference is applied, is $h_0(1 - \rho_0/\rho)$ where h_0 is the height of the lighter liquid (density ρ_0) above the interface with the heavier liquid.

4.4   Figure P4.4 shows a U-tube of cross-sectional area A which is sealed on the left-hand side, open on the right-hand side and contains a liquid of density ρ. The density of the gas above the liquid on the left-hand side is negligible. The solid cylinder of mass m on the right-hand side is completely supported by the liquid (i.e. the cylinder is a perfect fit in the tube with no leakage or friction). Derive an expression for the absolute pressure of the gas if the external pressure is B.

4.5   Figure P4.5 shows an inverted U-tube manometer used to measure the pressure p of a gas in a pipe. The U-tube contains two liquids, of densities

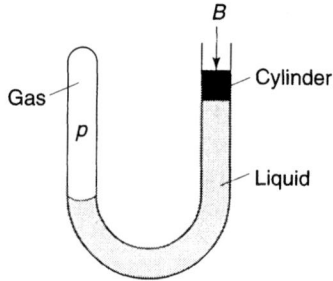

Figure P4.4

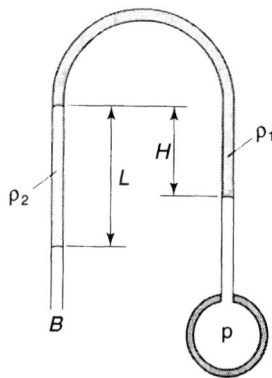

Figure P4.5

ρ_1 and ρ_2 as shown. Show that

$$p = B + g(\rho_1 H - \rho_2 L)$$

where B is the external pressure and g is the acceleration due to gravity. The gas density may be assumed to be negligible.

4.6 The pressure measured at the summit of a mountain is 0.31 bar. Calculate the height of the mountain assuming the temperature decreases linearly with altitude at a rate of 6.5°C/km. Take the pressure at sea level as 1.015 bar and the temperature as 15°C. What would be the error in calculating the height assuming that the air density was constant at its sea-level value? (Answers: 8947 m, −34.6%)

4.7 (a) Figure P4.7 shows a manometer consisting of two vertical arms of cross-sectional area a connected to form a U-tube. The reservoirs at the top of each arm are identical and of cross-sectional area A. The fluid in the right arm is water with density ρ_w, and the fluid in the left arm is an oil with density ρ_O which is less than ρ_W. The manometer is used to measure the difference between the pressures p_1 and p_2 for a gas of negligible density in contact with the liquids in the manometer, as shown in the figure. Show first that when p_1 and p_2 are equal, the difference in height between the oil and the water is

$$H_O\left(1 - \frac{\rho_O}{\rho_W}\right)$$

where H_O is the height of the oil column measured above the oil/water interface. Show further that a pressure difference $p_2 - p_1$ moves the interface an amount h given by

$$h = \frac{p_2 - p_1}{\rho_W g(1 + (a/A)) - \rho_O g(1 - (a/A))}$$

(b) If the oil has a density of 800 kg/m³ and the manometer tube an internal diameter of 5 mm, calculate the reservoir diameter required if the manometer reading for a pressure difference of 200 Pa is to be 10 cm. What would the manometer read for a pressure difference of 200 Pa if the reservoirs had the same internal diameter as the manometer tube? (Answers: 107.8 mm, 0.01 m)

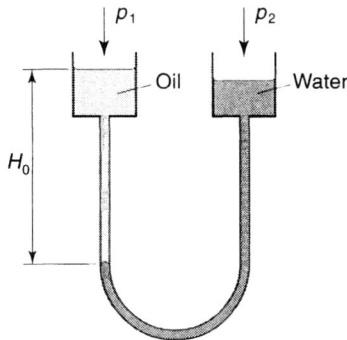

Figure P4.7

4.8   (a) The gas pressure p in a tank partially filled with oil is to be measured using a mercury manometer as shown in Figure P4.8. The oil and mercury are separated by a tube containing nitrogen. The manometer zero level corresponds to the situation when the nitrogen and the space above the oil are at the same pressure p_R (i.e. below the level in the mercury reservoir).

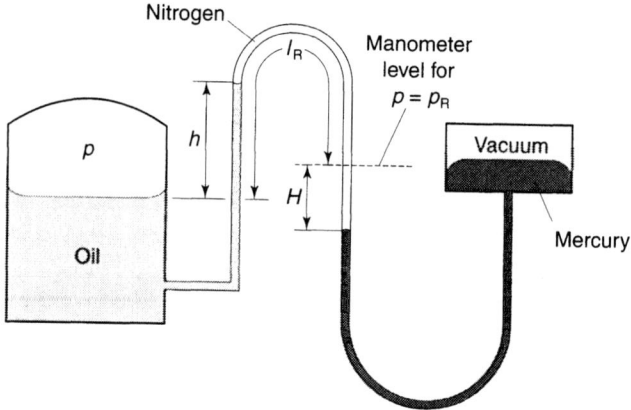

Figure P4.8

The length of manometer tubing occupied by the nitrogen is then l_R. If the density of nitrogen is negligibly small, show that p is given by

$$p - p_R = p_I + \rho_O gh$$

where

$$p_I = \rho_M(1 + r)gH \quad \text{and} \quad h = H + \left(\frac{p_I}{p_I + p_R}\right)l_R.$$

ρ_M and ρ_O are the densities of mercury and oil, respectively, g is the acceleration due to gravity, H is the change in the mercury level in the manometer tubing when p is different from p_R, and r is the ratio of the cross-sectional area of the manometer tubing to that of the mercury reservoir. Assume that the nitrogen temperature remains constant so that its volume is inversely proportional to its pressure. Assume also that the oil tank is so large that the oil level is not significantly changed by changing the pressure p.

(b) For a manometer reading H of 0.7 m, calculate the gas pressure p if the manometer tubing has an internal diameter of 5 mm and the reservoir internal diameter is 100 mm. The length l_R is 2 m when the pressure p_R is 10^5 Pa. The relative densities of oil and mercury are 0.8 and 13.6, respectively.

If the nitrogen were assumed to be incompressible, calculate the manometer reading for a gas pressure p of 0.5 bar.
(Answers: 2.067 bar, −0.353 m)

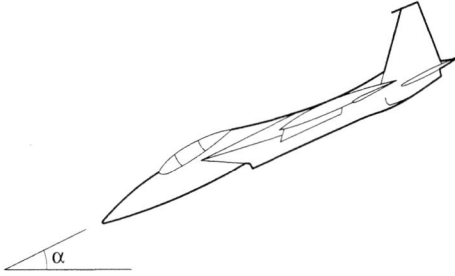

Figure P4.9

4.9   (a) A combat aircraft is flying in a nose-down attitude as shown in Figure
P4.9. If the aircraft is accelerating at a rate a, show that the isobars in the
fuel tanks are inclined at an angle θ to the horizontal given by

$$\tan \theta = \frac{a \cos \alpha}{g - a \sin \alpha}$$

where α is the pitch angle of the aircraft.
(b) If the aircraft in part (a) accelerates at $3g$ and pitch angle $30°$, calculate
the inclination of the isobars. If the aircraft now begins to climb with a
nose-up attitude of $30°$, calculate the inclination of the isobars if the
acceleration is again $3g$.
(Answers: $-79.1°$, $46.1°$)

CHAPTER 5

Hydrostatic force on a submerged surface

This chapter is concerned with *hydrostatic force*, which is the force exerted on a submerged body due to the hydrostatic pressure distributed over its surface. We start by showing that a completely uniform pressure acting on the surface of a solid object results in a zero net force. The remainder of the chapter is concerned with the forces exerted on submerged surfaces due to the linear increase with vertical depth of the pressure in a fluid of constant density. We show that the vertical component of the net hydrostatic force acting on a submerged surface is equal to the weight of fluid which occupies (or could occupy) the volume directly above the surface. It is also shown that the difference between the hydrostatic forces acting on the lower and the upper surfaces of a submerged body is the buoyancy force of Archimedes' principle. We then analyse the horizontal component of the hydrostatic force acting on a submerged surface. This component is shown to equal the hydrostatic force acting on an equivalent flat vertical surface but is less straightforward to calculate than the vertical component. The chapter concludes by considering the stability of a body either fully submerged or floating in a fluid.

5.1 Resultant force on a body due to uniform surface pressure

Figure 5.1 shows a body of arbitrary shape which is subjected to a uniform external pressure B acting on its surface. A tube of infinitesimal cross-section δA is shown passing through the body and intersecting its surface at points X and Y where the surface areas are δA_1 and δA_2. Since the body is of arbitrary shape, these surface elements will be at arbitrary orientations to the line XY. Although the argument here applies to any three-dimensional shape, it may be easier for the reader to imagine the body has a two-dimensional shape, i.e. we are looking at a cross-section which would be the same for any plane parallel to this page.

If the angles between the line XY and the surface normals at X and Y are θ_1 and θ_2, respectively, then

$$\delta A = \delta A_1 \cos \theta_1 = \delta A_2 \cos \theta_2,$$

i.e. the area δA corresponds to the projections of δA_1 and δA_2 onto a plane

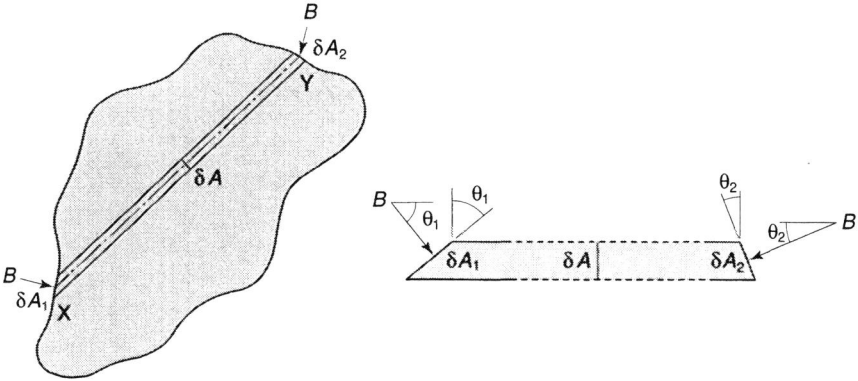

Figure 5.1 *Body of arbitrary shape surrounded by uniform pressure* B

perpendicular to the axis of the elemental tube. The external pressure B results in forces $B\ \delta A_1$ and $B\ \delta A_2$ acting normal to δA_1 and δA_2, respectively. The net force due to B acting along the line XY is thus

$$(B\ \delta A_1)\cos\theta_1 - (B\ \delta A_2)\cos\theta_2 = B(\delta A_1\cos\theta_1 - \delta A_2\cos\theta_2)$$

which must be zero because of the area relationship.

Since we can use the above argument for every part of the body surface, we can conclude that *a body of arbitrary shape subjected to a uniform external pressure experiences zero net force.*

This conclusion has an important consequence for the calculation of not only hydrostatic forces but also hydrodynamic forces (see Chapters 9 and 10): in situations where the pressure acting on a surface varies from point to point, we can add or subtract a uniform pressure everywhere without affecting either the hydrostatic or the hydrodynamic force balance. In particular, in problems for a liquid of constant density ρ, where the variation of the pressure p with depth z below the liquid surface is given by

$$p = B + \rho gz, \tag{4.6}$$

we can subtract from p the uniform barometric pressure B and *calculate hydrostatic forces using the hydrostatic or gauge pressure* $p_G = \rho gz$.

5.2 Vertical component of the hydrostatic force acting on a submerged surface

Just as in the case for solids, fluids have to obey Newton's laws of motion. From Newton's third law we can state: for a body of liquid at rest, the net force exerted

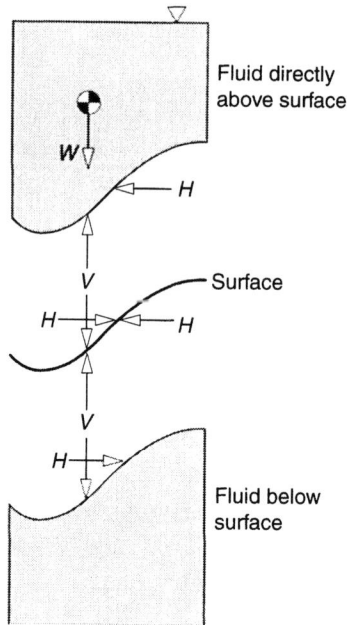

Figure 5.2 *Vertical component of the hydrostatic force acting on a surface submerged in a liquid*

by the liquid on any submerged solid surface must be equal in magnitude and opposite in direction to the force exerted by the surface on the liquid. Since we need to consider the hydrostatic forces on objects or structures which may have liquid above, below or both, the submerged surface in Figure 5.2 is shown as a thin weightless sheet with an upper and a lower surface.

The first condition which must be satisfied for the liquid directly above the sheet to be in static equilibrium is that there must be a vertically upwards force V exerted on the liquid by the upper surface of the sheet equal in magnitude to the weight of the fluid above it W, i.e.

$$W = V = \rho \vartheta g$$

where ρ is the liquid density, g is the acceleration due to gravity and ϑ is the volume of liquid directly above the sheet. *The vertical component of the hydrostatic force exerted by the fluid on the upper surface of the sheet must be of equal magnitude to the force exerted on the fluid* (i.e. $V = \rho \vartheta g$) *but act vertically downwards*, as indicated in Figure 5.2. Finally, for the lower surface of the sheet, the vertical component of the hydrostatic force must again be of magnitude W but act vertically upwards.

Each of the foregoing results could have been obtained by calculating the force due to the gauge pressure p_G distributed over the upper and lower surfaces of the sheet. In the case of the upper surface, the vertical component of the force would be obtained by integrating the vertical component of the force on every element of the surface δA, i.e.

$$\delta V = p_G \cos \theta \, \delta A = \rho g z \cos \theta \, \delta A$$

where z is the depth of a point on the sheet below the liquid surface and θ is the angle between the vertical and the normal to the surface of the sheet at that point. Since $\delta A \cos \theta$ is the area of an element of the sheet projected onto a horizontal plane, $\delta A \cos \theta \; z$ represents the volume $\delta \vartheta$ of the vertical cylinder of fluid directly above the surface element δA. The force V is thus

$$V = \int_\vartheta \rho g \, d\vartheta = \rho g \vartheta$$

as before. If the shape of the surface is simple (e.g. flat or cylindrical) there is a good chance the volume can be calculated easily from well-known formulae for the volumes of rectangles, prisms, cylinders, etc. However, as we shall show in a case study towards the end of this chapter, if the shape is more complex we have to evaluate the integral directly using the mathematical description of the shape.

So far we have considered only the magnitude and direction of V but not its line of action. The location of the latter is important because for a system of forces acting on a body to be in static equilibrium (i.e. for a stationary body to remain at rest) we also require that the forces exert no net moment on the body. As a consequence of this condition, for the liquid directly above the sheet *the line of action of V must pass vertically through the centroid of the liquid volume.*

We note that to calculate both the magnitude of V and location of its line of action, we are concerned essentially with the geometry (i.e. shape and size) of the volume ϑ.

EXAMPLE 5.1
Calculate the magnitude of the hydrostatic force exerted on the upper surface of the kite-shaped plate shown in Figure E5.1 and the location of its line of action if the plate is submerged horizontally at a depth Z below the surface of a liquid of density ρ.

SOLUTION
Since the plate is horizontal, the entire hydrostatic force acting on its surface must be vertical and act downwards through the centroid of the surface of the plate. The magnitude of the force V is equal to the weight of

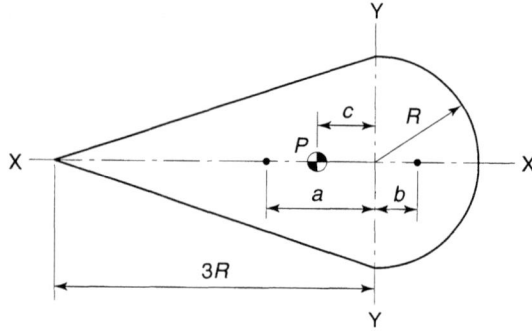

Figure E5.1

fluid directly above the plate, i.e.

$$V = \rho gAZ.$$

The surface area of the plate A is given by

$$A = 3R^2 + \tfrac{1}{2}\pi R^2 = (3 + \tfrac{1}{2}\pi)R^2$$

so that the magnitude of V is given by

$$V = (3 + \tfrac{1}{2}\pi)\rho gZR^2.$$

We can easily show that this result is consistent with what we obtain by considering the gauge pressure p_G acting on the surface of the plate. We note that because the plate is horizontal, p_G is constant:

$$p_G = \rho gZ$$

and the magnitude of V is given by

$$V = p_G A = \rho gZA,$$

exactly as before.

The symmetry of the kite about XX tells us that the line of action of V must pass through a point P somewhere along the line XX. The location of P is given by its distance c from the line YY which can be calculated by equating the moment of V about YY (or any other line parallel to YY) to the combined moments of the hydrostatic forces acting on the triangular section of the kite to the left of YY and on the semi-circular section to the right of YY:

$$\rho gZ.\,3R^2 a - \rho gZ.\tfrac{1}{2}\pi R^2 b = \rho gZ(3 + \tfrac{1}{2}\pi)R^2 c.$$

The locations of the lines of action of the hydrostatic forces acting on the triangular and semi-circular sections correspond with their centroids which we can find from Table A.9 as

$$a = R; \qquad b = \frac{4R}{3\pi}$$

so that, after dividing through by the common factor $\rho g Z R^2$, the moment equation becomes

$$3R - \frac{1}{2}\pi\frac{4R}{3\pi} = \left(3 + \frac{1}{2}\pi\right)c$$

from which

$$c = \frac{7R}{3(3 + \frac{1}{2}\pi)} = 0.51R.$$

Since the plate in this example is horizontal, the only difficulty in the problem stems from the shape of the plate. In the following example the surface shape is made up of two rectangles and the difficulty arises because they are sloping rather than horizontal. As we have already remarked, later in this chapter we shall deal with a situation where the surface shape is sufficiently complex that we need to resort to integration to find the volume of fluid directly above it.

EXAMPLE 5.2

A dam has the cross-section shown in Figure E5.2 with $\tan\theta = 4$. The water depth is H and the length of the dam is L. Calculate the vertical

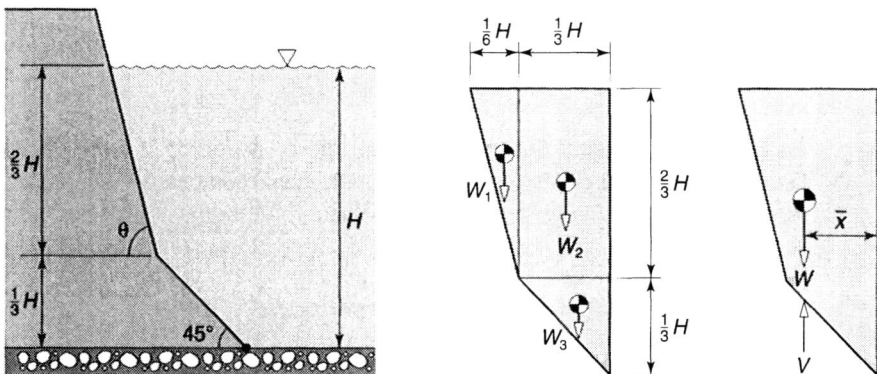

Figure E5.2

component of the hydrostatic force acting on the face of the dam and the horizontal distance of the line of action of this force from the point O.

SOLUTION

The vertical component V of the hydrostatic force acting on the face of the dam is given by

$$V = \rho \vartheta g$$

where ρ is the water density, g is the acceleration due to gravity and ϑ is the volume of water directly above the sloping surface.

As shown in the lower left-hand part of Figure E5.2, it is convenient to split ϑ into three smaller volumes for which the volumes and centroid locations can be found from Table A.9 as follows:

Volume 1: Cross-section is a right-angle triangle with

height $= \frac{2}{3} H$ base length $= \frac{1}{6} H$ (since $\tan \theta = 4$)

centroid $\frac{1}{18} H$ from vertical face   volume $\vartheta_1 = \frac{1}{2} \cdot \frac{1}{6} H \frac{2}{3} HL = \frac{1}{18} H^2 L.$

Volume 2: Cross-section is a rectangle with

height $= \frac{2}{3} H$ width $= \frac{1}{3} H$

centroid $\frac{1}{6} H$ from vertical face   volume $\vartheta_2 = \frac{1}{3} H \frac{2}{3} HL = \frac{2}{9} H^2 L.$

Volume 3: Cross-section is a right-angle triangle with

height $= \frac{1}{3} H$ base length $= \frac{1}{3} H$

centroid $\frac{1}{9} H$ from vertical face   volume $\vartheta_3 = \frac{1}{2} \cdot \frac{1}{3} H \frac{1}{3} HL = \frac{1}{18} H^2 L.$

From the above

$$\vartheta = \vartheta_1 + \vartheta_2 + \vartheta_3 = (\tfrac{1}{18} + \tfrac{2}{9} + \tfrac{1}{18})H^2 L = \tfrac{1}{3} H^2 L.$$

Hence

$$V = W = \rho \vartheta g = \tfrac{1}{3} \rho H^2 Lg.$$

To find the horizontal distance $\bar{x}$ of the line of action of V from O, we equate the moment of V about O to the combined moments of the vertical forces V_1, V_2 and V_3 due to the weight of the three liquid volumes ϑ_1, ϑ_2 and ϑ_3:

$$V\bar{x} = (\tfrac{1}{3} + \tfrac{1}{18})HV_1 + \tfrac{1}{6} HV_2 + \tfrac{1}{9} HV_3$$

$$= (\tfrac{1}{18} \cdot \tfrac{7}{18} + \tfrac{2}{9} \cdot \tfrac{1}{6} + \tfrac{1}{18} \cdot \tfrac{1}{9})\rho H^3 Lg$$

$$= \tfrac{7}{108} \rho H^3 Lg$$

from which, if we substitute $\frac{1}{3}\rho H^2 Lg$ for V, we have

$$\bar{x} = \frac{\frac{7}{108}\rho H^3 Lg}{\frac{1}{3}\rho H^2 Lg} = \frac{7}{36}H.$$

As we pointed out earlier in this section, the thin sheet depicted in Figure 5.2 has both an upper and a lower surface. For this weightless sheet to be in equilibrium, the net force in any direction must be zero. The hydrostatic force on the downward-facing surface must therefore be equal in magnitude and opposite in direction to the hydrostatic force on the upward-facing surface, and this must also apply to the vertical and horizontal components of the hydrostatic force. While this conclusion may seem to be quite obvious for a submerged sheet, it may be less apparent if there is liquid beneath the sheet but not above.

The shape of the surface shown in Figure 5.3 is the same as that of Figure 5.2 but the vertical sides now prevent contact between the liquid and the upper surface of the sheet. Since the hydrostatic pressure is constant along any horizontal line within a liquid at rest, the pressure distribution over the downward-facing surface is completely unaffected by the fact that there is no liquid above the sheet. The hydrostatic force must also be unaffected since it represents the integrated effect of the pressure distributed over a surface. We conclude that *the magnitude of the vertical component of the hydrostatic force exerted on a surface submerged in a liquid is equal to the weight of the liquid which occupies the volume directly above the surface or of the liquid which could occupy this volume.*

EXAMPLE 5.3

Figure E5.3 shows the cross-section of an axisymmetric container. The upper and lower sections are both cylindrical with radii r and R, respectively, and are separated by a conical section which slants at an angle θ to the horizontal. If the container is filled with a liquid of density ρ to a height

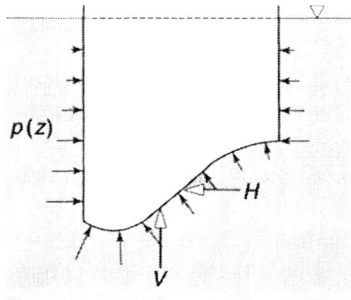

Figure 5.3 *Surface with liquid below but not above*

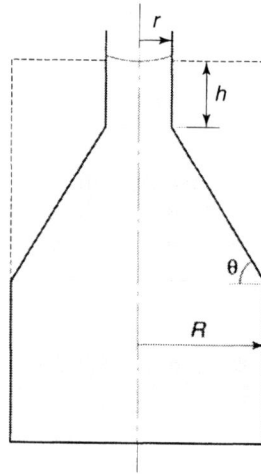

Figure E5.3

H above the top of the conical section, calculate the magnitude of the hydrostatic force exerted by the liquid on the conical section. State the direction of this force and the location of its line of action.

SOLUTION

Since the container is axially symmetric about its vertical axis, the radial (i.e. horizontal) component of force arising from the pressure acting on any element of surface area is counteracted by a force of equal magnitude but opposite direction arising on an identical element of area on the other side of the container. Due to the axisymmetry, therefore, the net hydrostatic force on the conical section of the container must be vertical and its line of action coincident with the axis of symmetry. From the geometry of the conical section it must also be that the hydrostatic force is directed vertically upwards.

The magnitude of the hydrostatic force is given by $\rho \vartheta g$, where ϑ is the volume of liquid which could occupy the space directly above the conical surface up to the level of the free surface of the liquid (shown in Figure E5.3 by the broken lines).

We can calculate ϑ by subtracting from the volume of a cylinder of radius R and height $h + (R - r) \tan \theta$ the volumes of the small cylindrical section (height h, radius r) and of the interior of the conical section (a frustum of a cone of upper radius r, lower radius R and slope angle θ), i.e.

$$\vartheta = \pi R^2[h + (R - r) \tan \theta] - \pi r^2 h - \tfrac{1}{3} \pi (R^3 - r^3) \tan \theta$$

5.3 Archimedes' principle and the buoyancy force on submerged bodies

A body of arbitrary shape submerged in a liquid, such as is shown in Figure 5.4, can be thought of as having an upper surface and a lower surface. From the previous section of this chapter, the vertical component of the hydrostatic force exerted on the upper surface V_U will be a downwards force of the same magnitude as the weight W_U of the liquid in the volume ϑ_U directly above the upper surface, i.e.

$$V_U = W_U = \rho \vartheta_U g.$$

Similarly, the vertical component of the hydrostatic force exerted on the lower surface V_L will be an upwards force of the same magnitude as the weight W_L of the liquid which could occupy the entire volume (i.e. including the body itself) ϑ_L directly above the lower surface, i.e.

$$V_L = W_L = \rho \vartheta_L g.$$

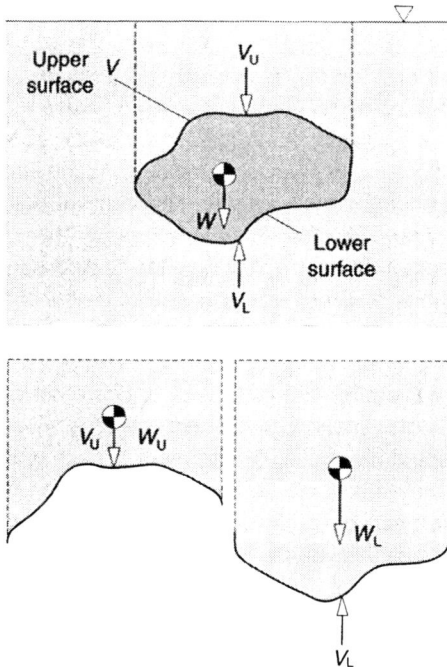

Figure 5.4 *Hydrostatic forces on a completely submerged body*

The magnitude of the net vertical component of the hydrostatic force acting on the body V_B must equal the difference between V_U and V_L, i.e.

$$V_B = V_L - V_U = \rho(\vartheta_L - \vartheta_U)g$$

Since the volume ϑ_L is made up of the volume ϑ_U together with the volume of the submerged body ϑ we have

$$\vartheta_L - \vartheta_U = \vartheta$$

and hence

$$\boxed{V_B = \rho\vartheta g.}\tag{5.1}$$

The subscript B has been introduced as a reminder that V_B is usually called the buoyancy force. As we can see from equation (5.1), *the magnitude of the buoyancy force is equal to the weight of liquid displaced by the submerged body*, a result known as Archimedes' principle after the Greek physicist and mathematician Archimedes (287–212 BC). According to legend this concept came to Archimedes while entering the pool of a public bath as the solution to the problem of determining the amount of gold used in the fabrication of the crown of the king of Syracuse.

Throughout most of this chapter we have used the word 'liquid' rather than 'fluid' because hydrostatics problems usually concern liquids rather than gases. In fact any result in this chapter which does not involve the depth below a free surface may be taken to apply to a surface or body immersed in any fluid. Equation (5.1), for example, can be used to calculate the buoyancy force exerted by the surrounding atmosphere on a lighter-than-air balloon.

Since $\vartheta_L > \vartheta_U$, it should be clear that the buoyancy force V_B always acts vertically upwards. This conclusion can also be seen to be a consequence of the increase in hydrostatic pressure with depth since the forces on the surface of a submerged body are due to the pressure distributed over it. It should also be apparent from section 5.2 that *the line of action of V_B must pass through the centroid of the displaced volume ϑ*. If the submerged body is of uniform density, the location of its centre of gravity will correspond with the centroid of the displaced liquid, but will not in general correspond if there is a variation of density within the interior of the submerged body, as would be the case for a submarine, for example.

If a body is submerged only partially rather than completely, Archimedes' principle is still valid but the volume ϑ must be taken as that part of the volume of the body which is below the surface of the liquid. Whether or not a body floats or sinks in a liquid is determined by the average density of the body $\rho_B = m/\vartheta$, where m is its mass. If $\rho_B > \rho$, the body will become completely submerged and sink (unless constrained) whereas if $\rho_B < \rho$, the body will float with part of its volume ($= \rho_B \vartheta/\rho$) below the surface and the rest above.

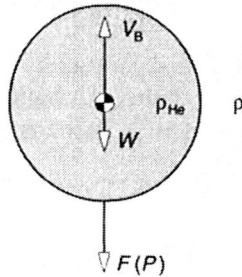

Figure E5.4

EXAMPLE 5.4

A rigid balloon in the form of a thin rigid sphere of diameter $D = 1$ m is filled with helium of density $\rho_{He} = 0.17$ kg/m³. Calculate the force required to prevent the balloon from rising if the surrounding air density at ground level $\rho = 1.2$ kg/m³ and the balloon material has negligible mass. What payload could the balloon lift to an altitude of about 8500 km where the ambient density is 0.5 kg/m³?

SOLUTION

The volume of the balloon ϑ is given by

$$\vartheta = \tfrac{1}{6}\, \pi D^3 = 0.52 \text{ m}^3.$$

The corresponding buoyancy force at ground level V_B is thus

$$V_B = \rho \vartheta g = 1.2 \times 0.52 \times 9.81 = 6.12 \text{ N}.$$

The weight of the filled balloon W is

$$W = \rho_{He} \vartheta g = 0.17 \times 0.52 \times 9.81 = 0.87 \text{ N}.$$

From Figure E5.4 we can see that static equilibrium at ground level requires

$$V_B - W - F = 0$$

so that the force F required to prevent the balloon from rising is

$$F = V_B - W = 5.25 \text{ N}.$$

If the balloon is released with a payload P less than F, it will rise to an altitude where

$$P = V_B - W.$$

Since the balloon is rigid, both its volume ϑ and weight W remain the same as at ground level. However, the buoyancy force V_B will decrease with altitude in direct proportion to the ambient density. For $\rho = 0.5$ kg/m^3 we have

$$V_B = 0.5 \times 0.52 \times 9.81 = 2.55 \text{ N}$$

and the payload P for static equilibrium is given by

$$V_B - W = 1.68 \text{ N}.$$

Although the pressure does not appear in equation (5.1), it is worth while reminding ourselves that the buoyancy force is a direct consequence of the increase in pressure with depth (or decrease with altitude), as illustrated by Figure 5.7. What is quite remarkable is that although in the atmosphere this pressure difference is usually very small, the resulting force can be quite substantial. For example, for a balloon of diameter 1 m the pressure difference Δp from top to bottom is only 0.012% of 1 bar, i.e.

$$\Delta p = \rho Dg = 1.2 \times 1 \times 9.81 = 11.8 \text{ Pa},$$

but, as we found, the lift force at ground level was 6.12 N, which is roughly equal to the weight of a pint of beer. The term *lift force* was used quite deliberately: just as for a balloon, the lift force on a wing arises because the average pressure on the lower surface is higher than that on the upper surface. As we shall show in section 8.5, the difference of course is that, in the case of a wing, the pressure difference is a consequence of its forward motion through the air, a much more complicated process than the hydrostatic pressure difference due to gravity.

Before we leave the topic of balloons, it is interesting to calculate the force required to submerge the helium-filled balloon of Example 5.4 in water. The buoyancy force V_B is now enormous

$$V_B = 10^3 \times 0.52 \times 9.81 = 5.1 \times 10^3 \text{ N}$$

and this is effectively the force which must be overcome in order to submerge the balloon since the weight (unchanged at 0.87 N) is obviously negligible.

5.4 Hydrostatic force on a submerged vertical surface

Figure 5.5 shows the front and side views of a flat surface of area A submerged vertically in a liquid of density ρ. To the right-hand side of the figure is a graph showing the proportional increase in hydrostatic pressure p_H with depth z, i.e.

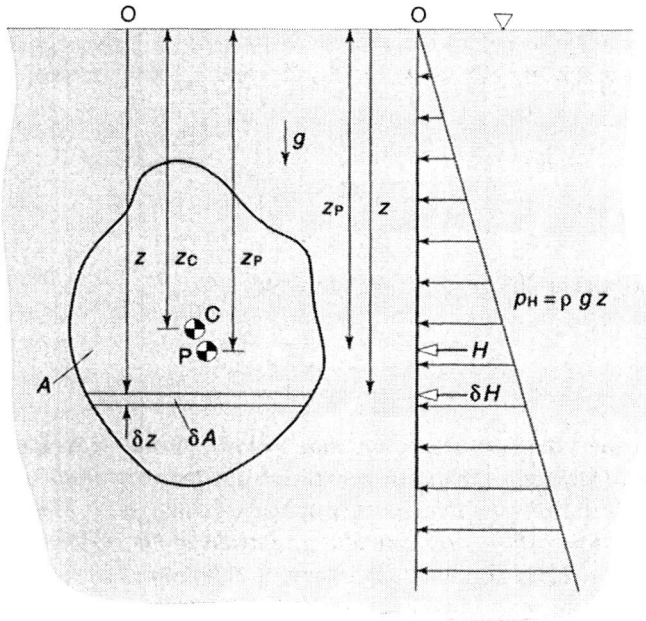

Figure 5.5 *Flat surface, submerged vertically*

$p_H = \rho gz$. Since this pressure acts normal to the surface at every point, the direction of the net hydrostatic force H exerted by the liquid on the surface must be horizontal.

To calculate the magnitude of H we split the area A into a series of infinitesimal horizontal strips, such as the strip of area δA at depth z shown in Figure 5.5. Since the hydrostatic pressure p_H is constant along a horizontal line, the infinitesimal hydrostatic force δH exerted on the strip is given by

$$\delta H = p_H\, \delta A = \rho gz\, \delta A$$

and the net force H is obtained by summing all such elemental forces over the area A, i.e.

$$H = \sum \delta H = \sum_A \rho gz\, \delta A$$

As usual in the limit as δA approaches zero, the summation can be replaced by an integration so that

$$H = \rho g \int_A z\, \mathrm{d}A.$$

Since ρ and g are constants, we regard them simply as multipliers of the integral $\int_A z\,dA$. This integral, called *the first moment of area*, turns out to be rather special because it is directly related to the location z_C of the centroid of the area A, as follows

$$\int_A z\,dA = z_C A.$$

We can make use of this general result to write

$$\boxed{H = \rho g z_C A = p_C A} \tag{5.2}$$

where p_C denotes the hydrostatic pressure at the centroid of A. Equation (5.2) represents an extremely important general result: *the magnitude of the hydrostatic force exerted on a vertical surface submerged in a fluid of uniform density is equal to the hydrostatic pressure at the centroid of the surface multiplied by its surface area*. The enormous advantage of equation (5.2) is that, for many standard shapes, the area A and location of the centroid are either well known to us (e.g. for the rectangle and the circle) or tabulated as in Table A.9, and more complicated shapes can often be treated as combinations of simpler ones, much like the kite shape of section 5.2. Of course, for non-standard or complicated shapes we have to evaluate the integral $\int_A z\,dA$ from first principles although, as we shall illustrate shortly with an example, this can be quite an effort. First, though, we need to determine the line of action of H, i.e. to calculate the depth z_P of the point P on the surface, known as the *centre of pressure*, where the force H can be considered to act (we should not lose sight of the fact that H is not a real force but a force which, so far as the static-equilibrium of the surface is concerned, could replace the pressure distributed over the surface A). The reader might ask why the point P is not coincident with the centroid C, as was the case for the vertical hydrostatic force acting on a flat horizontal surface. The answer is that this would be the case if the pressure acting on the plate were uniform and equal to the average hydrostatic pressure H/A, which is exactly equal to the hydrostatic pressure at the centroid p_C. But at every point above C the pressure is less than p_C, and at every point below C the pressure is greater than p_C, so the hydrostatic force must always act at a point below C.

Once we know the magnitude of any force, its line of action is always determined in the same way: by taking moments about a convenient point. In the present case we take a point O in the surface directly above the plane of the surface.

For the elemental force δH the moment about O is

$$\delta H\, z = p_H\, \delta A\, z = \rho g\, z^2\, \delta A$$

and the net moment of all such elements is given by

$$\int_A \rho g z^2 \, dA = H z_p = \rho g z_C A z_p$$

wherein we have equated the net moment due to all the elemental forces to the moment of the net force H acting at the depth z_p. If we cancel out the common factor ρg, we find

$$z_p = \frac{\displaystyle\int_A z^2 \, dA}{z_C A}$$

The integral $\int_A z^2 \, dA$ is again special and will be encountered by all students of mechanical, aeronautical, civil, structural engineering, etc., for example in the stress analysis of beams and dynamics of rotating objects: it is the second moment of the area A (second because the integral involves z^2) and closely related to the moment of inertia about a horizontal axis through O parallel to the plane of the plate. This integral is easily related to the second moment of area taken about a parallel axis passing through the centroid C, as follows.

Since the centroid is at depth z_C, we can write

$$z = z_C + y$$

where (as shown in Figure 5.5) y is the vertical depth of the elemental strip below the centroid. The integral can now be written as follows

$$\int_A z^2 \, dA = \int_A (z_C^2 + 2 z_C y + y^2) \, dA$$

$$= z_C^2 A + 2 z_C \int_A y \, dA + \int_A y^2 \, dA.$$

The integral $\int_A y \, dA$ can be shown to be zero as follows

$$\int_A z \, dA = \int_A (z_C + y) \, dA = z_C A + \int_A y \, dA$$

but, as we have just seen, z_C is defined by $\int_A z \, dA = z_C A$, so that

$$\int_A y \, dA = 0 \quad \text{and} \quad \int_A z^2 \, dA = z_C^2 A + \int_A y^2 \, dA.$$

We then find

$$z_P = \frac{\displaystyle\int_A z^2 \, dA}{z_C A} = z_C + \frac{\displaystyle\int_A y^2 \, dA}{z_C A}$$

$$= z_C + \frac{I_C}{z_C A}. \tag{5.3}$$

The result just obtained confirms what we argued above: P is always below C because $y^2 > 0$ and so $z_P - z_C > 0$. The integral $\int_A y^2 \, dA = I_C$ is called *the second moment of area* about an axis through the centroid and is again a tabulated quantity. It is important to realise that the value of the second moment of area depends upon the orientation of the axis. Table A.9 gives the areas, centroids and second moments of area for a number of shapes commonly encountered in engineering problems.

EXAMPLE 5.5
A circular disk of radius R is immersed vertically in a liquid of density ρ with its centre a distance Z below the surface. Calculate the hydrostatic force H which the liquid exerts on one face of the disk and the depth Z_P at which it acts.

SOLUTION
The surface area A of the disk is given by

$$A = \pi R^2$$

and the centroid is coincident with its centre so that

$$z_C = Z.$$

The hydrostatic pressure at the centroid p_C is thus

$$p_C = \rho g Z$$

and from equation (5.2) the hydrostatic force H exerted by the liquid on the disk is given by

$$H = p_C A = \rho g Z \pi R^2.$$

From equation (5.3), the depth at which H acts is given by

$$z_P = Z + \frac{I_C}{\pi R^2 Z}.$$

From Table A.9 the second moment of area about the centroid of the disk is

$$I_C = \frac{\pi R^4}{4}$$

so that

$$z_P = Z + \frac{R^2}{4Z}.$$

Little could be more straightforward than the above. We suppose now that we do not know the area of the disk, the location of its centroid or its second moment of area so that it is necessary to calculate H and z_P from first principles.

The disk is shown in Figure E5.5. We start by identifying a horizontal strip of infinitesimal width δz somewhere on the surface of the disk at depth z. It is convenient, as shown, to work in cylindrical coordinates, i.e.

$$z = Z - R \cos \theta \quad \text{and} \quad \delta z = R \sin \theta\, \delta\theta.$$

The area of the strip δA is then given by

$$\delta A = 2R \sin \theta\, \delta z = 2R^2 \sin^2 \theta\, \delta\theta.$$

The hydrostatic pressure p_H at depth z is given by

$$p_H = \rho g z = \rho g(Z - R \cos \theta)$$

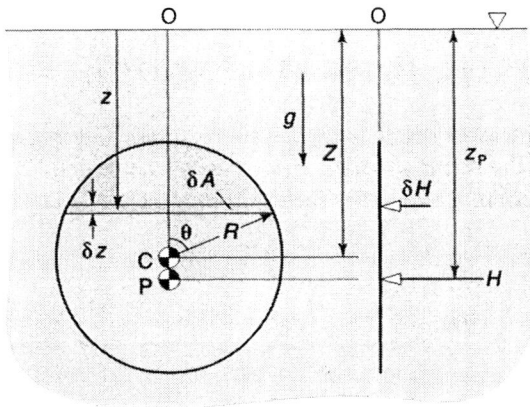

Figure E5.5

so that the elemental hydrostatic force acting on the strip δH is

$$\delta H = p_H \, \delta A = \rho g (Z - R \cos \theta).2R^2 \sin^2 \theta \, \delta \theta.$$

The hydrostatic force H is then calculated from

$$H = \rho g.2R^2 \int_0^\pi (Z - R \cos \theta) \sin^2 \theta \, d\theta$$

which we can split into two separate parts as follows

$$H = \rho g.2R^2 \left(Z \int_0^\pi \sin^2 \theta \, d\theta - R \int_0^\pi \cos \theta \sin^2 \theta \, d\theta \right).$$

The second of the two integrals is found from

$$\int_0^\pi \cos \theta \sin^2 \theta \, d\theta = \int_0^0 \sin^2 \theta \, d(\sin \theta) = 0.$$

The first integral is trickier but can be shown to be (either using integration by parts or from tables)

$$\int_0^\pi \sin^2 \theta \, d\theta = \tfrac{1}{2} \pi.$$

Finally we have

$$H = \rho g Z.2R^2 . \tfrac{1}{2} \pi = \rho g Z \pi R^2,$$

which is the same result as before but required a lot more effort to obtain.

To find the depth at which H acts from first principles, we need to take moments about a line in the surface through O and parallel to the plane of the disk:

$$Hz_P = \int_{z=Z-R}^{z=Z+R} z \, dH$$

$$= 2\rho g R^2 \int_0^\pi (Z - R \cos \theta)^2 \sin^2 \theta \, d\theta.$$

Since $H = \rho g Z \pi R^2$, we can re-write this equation as follows

$$\frac{\pi z_P}{2Z} = \int_0^\pi \left(1 - \frac{R \cos \theta}{Z}\right)^2 \sin^2 \theta \, d\theta$$

$$= \int_0^\pi \left[\sin^2 \theta - 2\left(\frac{R}{Z}\right)\cos \theta \sin^2 \theta + \left(\frac{R}{Z}\right)^2 \cos^2 \theta \sin^2 \theta\right] d\theta.$$

As before,

$$\int_0^\pi \sin^2 \theta \, d\theta = \tfrac{1}{2}\pi,$$

$$\int_0^\pi \cos \theta \sin^2 \theta \, d\theta = 0,$$

and from tables

$$\int_0^\pi \cos^2 \theta \sin^2 \theta \, d\theta = \tfrac{1}{8}\pi.$$

The equation for z_P then becomes

$$\frac{\pi z_P}{2Z} = \tfrac{1}{2}\pi + \left(\frac{R}{Z}\right)^2 \tfrac{1}{8}\pi \quad \text{or} \quad z_P = Z + \frac{R^2}{4Z},$$

once again the same result as before.

5.5 Hydrostatic force acting on a curved surface

In this section we show how the results obtained so far can be used to calculate the magnitude and direction of the horizontal and vertical components, H and V, of the resultant hydrostatic force R acting on a submerged surface of any shape. To illustrate the general approach, for convenience we use the example of a dam with the cross-section shown in Figure 5.6. The span of the dam is S and the particular shape shown (i.e. the curve representing the surface in contact with the water) is given by

$$y = \frac{Cx^2}{D}$$

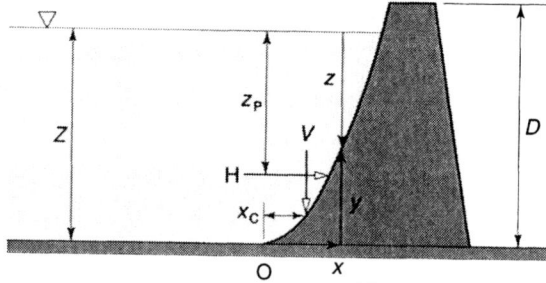

Figure 5.6 *Hydrostatic forces acting on the face of a dam*

where C is a constant, y is the vertical distance from the foot of the dam, x the corresponding horizontal distance, and D represents the vertical height of the dam. In the analysis below it is also convenient to introduce z, the depth below the surface, i.e.

$$z = Z - y$$

where Z is the total water depth.

Also shown in Figure 5.6 is the line of action of H at depth z_P and the line of action of V a horizontal distance x_C from the foot of the dam O.

5.5.1 *Horizontal component of* R

We consider an elemental horizontal strip of the surface of the dam a depth z below the water surface, as shown in Figure 5.7. The hydrostatic pressure p_H acting on the strip is given by

$$p_H = \rho g z.$$

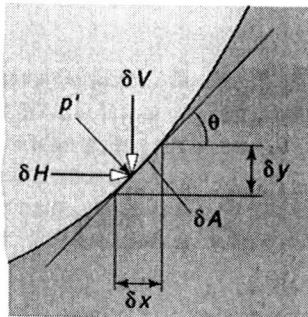

Figure 5.7 *Hydrostatic forces on an elemental strip*

The strip is assumed to be so narrow that it may be considered to be flat. The elemental horizontal force δH acting on the strip is then given by

$$\delta H = p_H \, \delta A \sin \theta$$

where δA is the area of the strip and θ the angle between a tangent to the strip and the horizontal. We can see from Figure 5.7 that

$$\delta A \sin \theta = \delta y \, S$$

so that

$$\delta H = \rho g z S \, \delta y$$

and the horizontal component of R is given by

$$H = \rho g S \int_0^Z z \, dy$$

$$= \rho g S \int_0^Z (Z - y) \, dy$$

$$= \tfrac{1}{2} \rho g S Z^2.$$

This result reveals that for the horizontal component of R, the cross-sectional shape of the dam is irrelevant and the magnitude of H is identical to the hydrostatic force exerted on a submerged vertical flat surface of area SZ. The area SZ represents the area of the shape obtained by projecting the dam onto a vertical plane. What we have is a general result: *the magnitude of the horizontal component of the hydrostatic force acting on a curved surface submerged in a fluid of uniform density is equal to the hydrostatic force exerted on a vertical flat surface, of area equal to the projected area of the curved surface, immersed in the same fluid.*

5.5.2 *Line of action of* H

As always we calculate the location of the line of action of H by taking moments about a suitable point. In this case we select the point O at the foot of the dam. The moment of the elemental force δH about O is given by $\delta H \, y$ where (from the preceding subsection) $\delta H = \rho g z S \, \delta y$. The net moment of all such elemental forces is then $\int_{y=0}^{y=Z} y \, dH$ which must equal the moment of H itself, i.e. since the line of action of H is at depth z_P,

$$H(Z - z_P) = \int_{y=0}^{y=Z} y \, dH$$

$$= \int_0^Z y\rho gzS \, dy$$

$$= \rho gS \int_0^Z y(Z-y) \, dy$$

$$= \tfrac{1}{6} \rho gSZ^3.$$

From subsection 5.5.1, we have $H = \tfrac{1}{2} \rho gSZ^2$ so that $z_P = \tfrac{2}{3} Z$.

Once again we see that only the water depth Z is important and the cross-sectional shape of the dam is of no consequence. This, too, is a general result: *the location of the line of action of the horizontal component of the hydrostatic force exerted on a curved surface submerged in a fluid of uniform density is the same as the location of the line of action of the hydrostatic force acting on a flat vertical surface, of area equal to the projected area of the curved surface, immersed in the same fluid.*

5.5.3 *Vertical component of* R

We have already shown in section 5.2 that the magnitude of the vertical component of the hydrostatic force exerted on a submerged surface is equal to the weight of the fluid which occupies the volume directly above the surface, i.e.

$$V = \rho g \vartheta.$$

We consider a vertical slice of the liquid directly above the elemental horizontal strip of the curved surface of the dam, of thickness δx, depth z and span S. The volume of the elemental slice is given by

$$\delta \vartheta = zS \, \delta x$$

so that the entire volume ϑ is

$$\vartheta = S \int_0^X z \, dx$$

where the symbol X has been introduced to denote the horizontal distance between the point O and the point where the water surface meets the curved face of the dam. We now have

$$V = \rho gS \int_0^X z \, dx$$

$$= \rho gS \int_0^X (Z-y) \, dx$$

and at this stage need to connect y and x by introducing the shape of the curved surface, i.e.

$$y = \frac{Cx^2}{D}.$$

The final result is

$$V = \tfrac{2}{3}\,\rho g S Z X$$

wherein we have also substituted $Z = CX^2/D$. We could have written down this result without further calculation had we known in advance (e.g. by reference to Table A.9) that the volume of liquid above a curved surface of parabolic cross-section is given by

$$\vartheta = \tfrac{2}{3}\, SZX.$$

5.5.4 *Line of action of* V

As we did for H, we calculate the location of the line of action of V by taking moments about O. We start with

$$V x_C = \int_{x=0}^{x=X} x\, \mathrm{d}V = \int_{x=0}^{x=X} \rho g x\, \mathrm{d}\vartheta$$

where x_C is the horizontal distance of the line of action of V from O and x is the horizontal distance from O of the elemental vertical slice of liquid.

We substitute $(Z - y)S\, \mathrm{d}x$ for $\mathrm{d}\vartheta$ and introduce $y = Cx^2/D$ for the shape of the dam to find

$$V x_C = \rho g S \left(\tfrac{1}{2} Z X^2 - \frac{C X^4}{4D} \right).$$

The final result

$$x_C = \tfrac{3}{8} X$$

is obtained after substituting $V = \tfrac{2}{3}\,\rho g S Z X$ and $Z = CX^2/D$.

As pointed out in section 5.2, x_C corresponds with the location of the centroid of the liquid directly above the curved surface.

5.5.5 *Resultant hydrostatic force* R

The resultant hydrostatic force R is calculated as the vector sum of H and V, i.e.

$$R = \sqrt{(H^2 + V^2)}$$

while the angle between R and the horizontal is given by $\tan^{-1}(V/H)$.

5.6 **Stability of a fully submerged body**

To introduce the concept of stability, we consider the behaviour of a simple pendulum which consists of a bob of weight W at the end of a weightless rod, of length l, supported by a pivot and free to swing in a vertical plane. The situation is illustrated in Figure 5.8. In position (a), where the centre of gravity of the bob G is vertically below the pivot O, both the net force and the moments acting on the pendulum equate to zero and the pendulum is at rest in a state of static equilibrium. If given an angular displacement θ to position (b) and then released, the pendulum will oscillate about, and eventually come to rest in, position (a), the oscillation being damped by friction at the pivot and the resistance to motion due to the fluid surrounding the pendulum. As is well known, in the absence of damping, for small displacements of a simple pendulum the oscillation is a simple harmonic motion with period $T = 2\pi\sqrt{(l/g)}$ where g is the acceleration due to gravity. It can be seen from Figure 5.8(b) that the motion of the pendulum is driven by the clockwise moment $Wl \sin\theta$ which always acts to decrease θ. *An object which returns to a position of static equilibrium when displaced from it is said to be in* stable equilibrium.

The position of the pendulum depicted in Figure 5.8(c) is also one of static equilibrium. However this position is unstable because the moment $Wl \sin\theta$ is now such that the response of the pendulum to the slightest displacement, as shown in Figure 5.8(d), is for the displacement θ to increase. The pendulum eventually moves to a position which is one of both static and stable equilibrium, in this case again position (a).

The close (but not perfect) analogy between the stability of a body completely

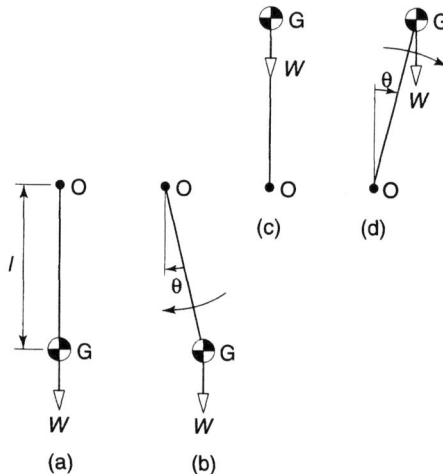

Figure 5.8 *Positions of a simple pendulum*

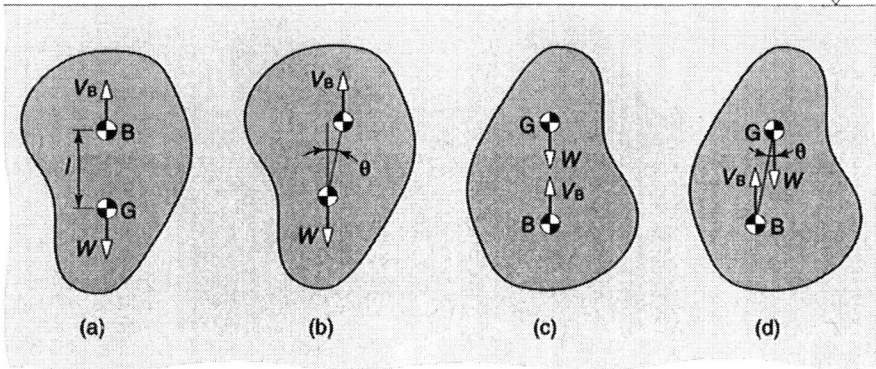

Figure 5.9 *Stability of a submerged body*

submerged in a fluid and that of a simple pendulum is illustrated by Figure 5.9: position (a) is again stable, position (c) is unstable, and both are positions of static equilibrium. The condition for static equilibrium now is that the centre of gravity of the body G is directly above or below the centre of buoyancy B (i.e. the centroid of the body). For a body of uniform density, B and G are coincident and the body is stable in any orientation. If B and G are not coincident, however, the angular displacement shown in Figures 5.9(b) and 5.9(d) gives rise to a moment $V_B l \sin \theta$, where V_B is the buoyancy force and l is the distance from the centroid to the centre of gravity. The analogy with the simple pendulum is not perfect because, when displaced, the body tends to roll about a horizontal axis close to its centre of gravity rather than a fixed pivot.

5.7 Stability of a freely floating body and metacentric height

The stability of a body which is partially submerged in a liquid (i.e. floating) is more complicated than for one which is completely submerged. The reason is that for a *completely submerged body the centre of buoyancy coincides with the centroid of the body* itself and this is always in the same place relative to the body. For a freely floating body the buoyancy force V_B must still equal the body's weight W and so the volume of displaced liquid remains constant (equal to $W/\rho g$). However, as illustrated in Figure 5.10, as the position of the body changes, for example due to a rolling motion, so does the shape of the submerged volume. In consequence, the centroid of the submerged volume, which defines the centre of buoyancy, is not fixed but is dependent on the body position, and this in turn has a critical influence on the stability of the floating body.

The upper part of Figure 5.10 shows a body of density ρ and weight W with its

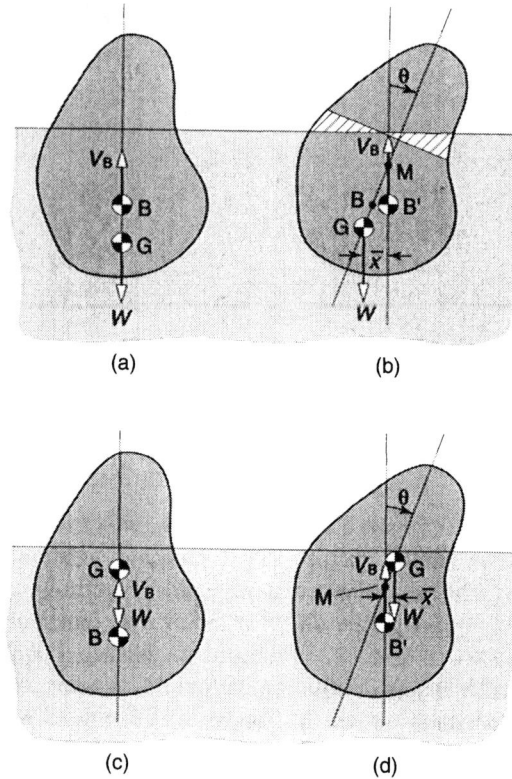

Figure 5.10 *Stability of a floating body*

centre of gravity G below the centre of buoyancy B of the submerged volume. When displaced through an angle θ from the vertical, the magnitude of the submerged volume ϑ must stay the same ($\vartheta = W/\rho g$) but its centroid will move a horizontal distance $\bar{x}$ with respect to G, to a new location B'. In this case, the couple $W\bar{x}$ exerted by the weight W and the buoyancy force $V_B(=W)$ acts to restore the body to the upright position: situation (a) is stable. The distance $\bar{x}$ is given by MG sin θ, MG being the distance between G and the point M where the line of action of the buoyancy force V_B intersects the line through BG. The point M is called the *metacentre* and the *length* MG *the metacentric height*, taken to be positive if, as in Figure 5.10(b), M is above G. For small angular displacements θ, MG is independent of θ, i.e. the location of M is fixed.

The lower part of Figure 5.10 shows a body of the same shape and weight as that in Figures 5.10(a) and (b). The centre of buoyancy B of the submerged volume must be in the same location as before but if the internal distribution of mass within the body is now different, the centre of gravity G of the body may now be above B. In the case of a ship, for example, the centre of gravity depends

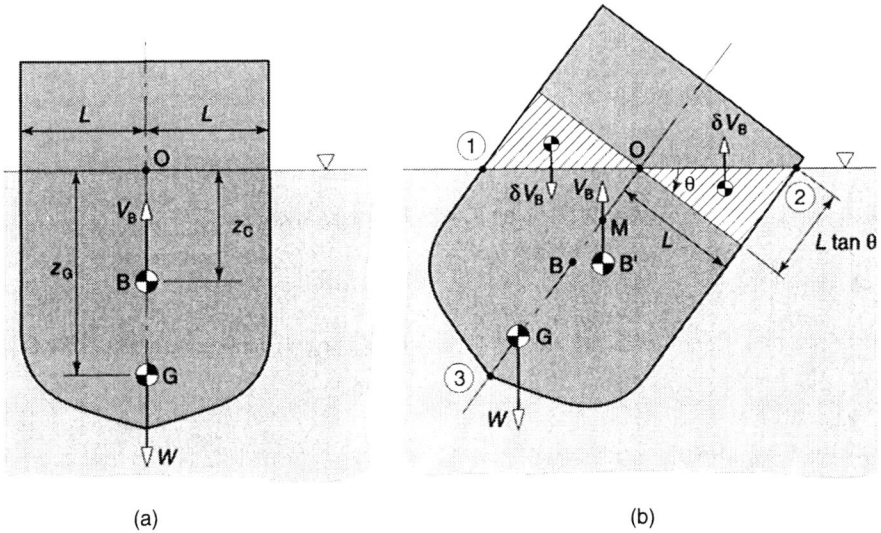

(a)   (b)

Figure 5.11 *Stability of a ship*

upon the way in which the cargo is distributed. As can be seen in Figure 5.10(d), the couple $W\bar{x}$ now acts to increase the angular displacement θ and so situation (c) is unstable. We can also see that the metacentre M is now below G and MG < 0. The magnitude and sign of MG have a critical influence on the stability of a floating body. If MG > 0 the body is in stable equilibrium, and the larger MG the more stable the situation. Exactly the opposite applies if MG < 0.

At first sight it appears that the condition for stability of a floating body is no different from that for one which is completely submerged, i.e. stable if G is below B, unstable if B is below G. In fact, as we now show, in certain circumstances a floating body with G slightly above B can be stable.

We consider the stability of a ship of length H with the cross-section shown in Figure 5.11(a). The depths below the surface of the centres of buoyancy B and gravity G are z_C and z_G, respectively. Figure 5.11(b) shows the situation if the ship is given an angular displacement θ.[*]

The new position of the centre of buoyancy B′ is obtained by calculating the centroid of the submerged volume defined by the shape ①, ②, ③. The metacentre M is defined by the intersection of the vertical through B′ and the line of symmetry OG. For a small displacement the centroid of the shaded triangular volume to the right of O is approximately a distance $\frac{2}{3}L$ from O. The buoyancy force δV_B corresponding to this volume can be taken as $\delta V_B = \frac{1}{2}L^2 \tan\theta\, H\rho g \approx \frac{1}{2}L^2\theta H\rho g$ if θ (measured in radians) is small.

---

[*] For a stability calculation we need consider only a small displacement. The angle θ shown in Figure 5.11(b) is greatly exaggerated.

Taking moments about O we see that

$$V_B.OM.\theta = V_B.OB.\theta - 2\,\delta V_B \tfrac{2}{3}\,L.$$

We have $V_B = W$, $OB = z_C$, and $OM = z_G - MG$ so that

$$W(z_G - MG) = Wz_C - \tfrac{2}{3}\rho L^3 Hg$$

or

$$MG = z_G - z_C + \frac{2\rho L^3 Hg}{3W}.$$

As we have already established, the ship will be stable if $MG > 0$ so that even if G is above B (i.e. $z_G < z_C$), stability is still possible if

$$z_C - z_G < \frac{2\rho L^3 Hg}{3W}.$$

EXAMPLE 5.6

A rectangular block of material has a width $2L$, height Z and length H. If σ is the relative density of the block ($\sigma < 1$), show that the block floats in water with its centre of gravity a distance z_G below the surface, given by

$$z_G = (\sigma - \tfrac{1}{2})Z.$$

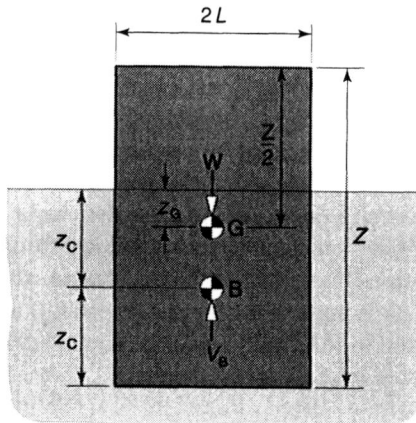

Figure E5.6

Show that the metacentre M is a distance MG above the centre of gravity G given by

$$MG = \tfrac{1}{2}(1 - \sigma)Z + \frac{2L^2}{\sigma Z}.$$

For given values of L and Z, what is the minimum value of σ for stability?

SOLUTION

The weight W of the block is given by

$$W = \rho.2LZHg$$

and the buoyancy force V_B is

$$V_B = \rho_F.2L.2z_C Hg$$

where ρ is the block density and ρ_F is the water density, i.e.

$$\rho = \sigma \rho_F.$$

The condition for static equilibrium must be satisfied if the block is floating freely, i.e.

$$W - V_B = 0 \quad \text{or} \quad \sigma \rho_F.2LZHg = \rho_F.2L.2z_C Hg$$

from which

$$z_C = \tfrac{1}{2}\,\sigma Z$$

We can see from Figure E5.6 that

$$\tfrac{1}{2}Z + z_G = 2z_C$$

so that

$$z_G = (\sigma - \tfrac{1}{2})Z.$$

From the ship example, we know that stability requires

$$z_C - z_G < \frac{2\rho_F L^3 Hg}{3W}$$

so that in this case

$$\frac{\sigma Z}{2} - (\sigma - \tfrac{1}{2})Z < \frac{L^2}{3\sigma Z} \quad \text{or} \quad \sqrt{[\tfrac{3}{2}\,\sigma(1-\sigma)]} < \frac{L}{Z}.$$

Comment: We see that a cube (i.e. $2L = Z$) will float upright if

$$\sqrt{[\tfrac{3}{2}\,\sigma(1-\sigma)]} < \tfrac{1}{2} \quad \text{or} \quad 6\sigma(1-\sigma) < 1$$

which leads to $\sigma > 0.79$ or $\sigma < 0.21$, i.e. a styrofoam cube ($\sigma < 0.2$) will float upright, but a cube of wood ($0.5 < \sigma < 0.7$) will not.

5.8 Summary

In this chapter we have calculated the force which arises due to the hydrostatic pressure exerted on a surface or object submerged in a fluid. For convenience we resolved the net force exerted on a surface into a vertical and a horizontal component. The vertical component was shown to be equal in magnitude to the weight of fluid which would occupy the volume directly above the surface and to act through the centroid of this volume. The buoyancy force exerted on a submerged or floating object was found to equal the weight of the fluid displaced by the object and to act vertically upwards through the centroid of displaced fluid. The relative positions of this centroid and the centre of gravity of the object were shown to determine its stability. We showed that for a flat surface immersed vertically in a fluid, the magnitude of the net hydrostatic force is equal to the product of the surface area and the pressure at the centroid of the surface. Because the hydrostatic pressure increases with depth, the line of action of this force always lies below the centroid. For a curved surface, the magnitude of the horizontal component of the hydrostatic force was shown to equal the hydrostatic force on the projection of the curved surface onto a vertical plane.

The student should be able to:

- calculate, both from first principles and from tabulated information for the properties of standard shapes, the magnitude and location of the line of action, and also specify the direction, of:
  – the vertical component of the hydrostatic force exerted on a submerged surface
  – the horizontal component of the hydrostatic force exerted on a submerged surface
  – the buoyancy force exerted on a submerged or floating object.
  The student should also be able to:
- analyse the stability of submerged or floating objects.

5.9 Self-assessment problems

5.1   The centroid of a vertical flat surface of area 0.5 m^2 completely submerged in an oil of relative density 0.85 is 5 m below the surface of the oil. Calculate the hydrostatic force acting on the surface. Explain why the hydrostatic force always acts some distance below the centroid. (Answer: 20.8 kN)

5.2   An aperture in the vertical wall of a water tank is closed by a circular plate 60 cm in diameter. The plate is held in position by four stops, one at each end of the horizontal diameter and one at each lower end of the diameters at 60° to the horizontal. Determine the stop reactions when the water surface is 45 cm above the plate centre. (Answers: 504.0 N, 120.1 N)

5.3   (a) The spread of an oil slick of depth d and density ρ_O is to be stopped by a floating boom (see Figure 1.5). The boom has a square cross-section of side t, a weight per unit length w and is designed to float upright (i.e. its sides are always vertical). Show that the maximum slick depth Z which can be contained by the boom is given by

$$Z = \left[\frac{1 - (w/\rho_S g t^2)}{1 - (\rho_O/\rho_S)} \right] t$$

where ρ_S is the density of the sea water beneath the slick and g is the acceleration due to gravity.

(b) Calculate the horizontal force per unit length acting on the water side of a boom if t is 500 mm, w is 2300 N/m and the relative density of sea water is 1.025. Calculate also the depth below the top face of the boom at which the horizontal force acts. For a rectangle with sides of length b and d, the second moment of area about an axis passing through the centroid and parallel to the side of length b is $bd^3/12$.

Would the horizontal force on the oil side of the boom be greater or smaller than that on the water side? Give a brief explanation for your answer.

(Answers: 1052 N/m, 0.348 m)

5.4   (a) A square plate of side length L is submerged in water at angle θ to the vertical with its centroid a depth Z below the surface and two sides parallel to the surface. Show that the net hydrostatic force acting on one face of the plate acts at a depth Z_P given by

$$Z_P = \frac{z_2^3 - z_1^3}{3ZL \cos \theta}$$

where $z_2 = Z + \frac{1}{2} L \cos \theta$ and $z_1 = Z - \frac{1}{2} L \cos \theta$.

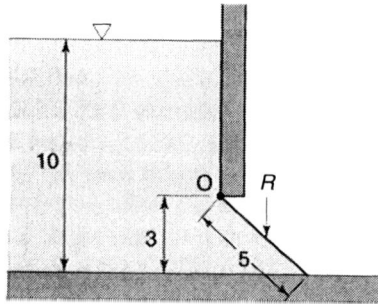

Figure P5.4

(b) The gate shown (side view) in Figure P5.4 is hinged at O which is 3 m above the bed of a reservoir which contains water of depth 10 m. If the gate is a square of side length 5 m, calculate the force R applied vertically downwards at its centroid necessary to prevent the gate from opening. Neglect the weight of the gate and any effects of leakage under the gate. There is no water on the right-hand side of the gate. (Answer: 2.76 MN)

5.5 (a) A dam has the cross-section shown in Figure P5.5 with $\tan \theta = 4$. Show that the resultant hydrostatic force R acting on the dam is given by

$$R = \frac{\sqrt{13}}{6} \rho g H^2 W$$

where H is the total water depth, as shown, ρ is the water density, g the acceleration due to gravity and W the width of the dam.

Show also that the horizontal distance X from 0 of the line of action of

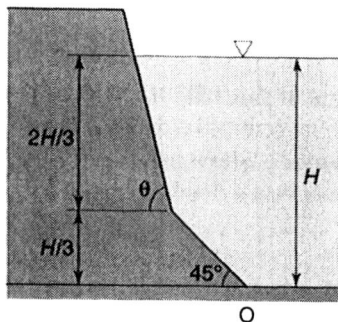

Figure P5.5

the vertical component of the hydrostatic force acting on the dam is given by $7H/36$. The centroid of a triangle is located at one-third of the height of any apex above the corresponding side.

(b) Calculate separately the horizontal and vertical components of the hydrostatic force on a dam of width 600 m if the water depth is 30 m. What angle does the resultant force make with the vertical and at what depth does the horizontal component of the hydrostatic force act?

(Answers: 2.65 GN, 1.77 GN, 56.3°, 20 m)

5.6 (a) The water in a tank of depth D and width W is prevented from escaping by a gate of circular cross-section (Figure P5.6) hinged at 0. If the radius of the gate is $D/2$, show that the net hydrostatic force F acting on the gate surface is given by

$$F = \tfrac{1}{4} \rho g W D^2 \sqrt{(\tfrac{25}{4} - \pi + \tfrac{1}{16} \pi^2)}$$

where ρ is the water density and g the acceleration due to gravity. There is no water to the left of the gate.

Show also that the vertical component of F acts at a horizontal distance

$$X = \frac{D}{6(1 - \tfrac{1}{8} \pi)}$$

from the tank wall. The centroid of a quadrant of radius R is a distance $4R/3\pi$ from either of the straight sides.

(b) If the depth D is 10 m and the width W is 5 m, calculate the force which the gate exerts on the tank wall. The weight of the gate can be neglected.

(Answer: 1.29 MN)

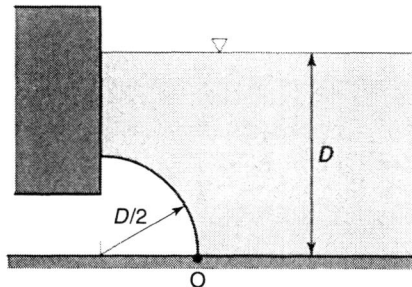

Figure P5.6

5.7 (a) Due to increasing salt levels and the presence of silt, the density ρ of water in a reservoir increases with depth z below the water surface according to

$$\rho = \rho_0 + Cz$$

where ρ_0 is the density at the surface ($z = 0$) and C is a constant. Show from first principles that the hydrostatic pressure at depth z is given by

$$p = (\rho_0 + \tfrac{1}{2} Cz)gz$$

and that the hydrostatic force H on a vertical rectangular wall of width w due to water of depth D is given by

$$H = \tfrac{1}{2} wgD^2(\rho_0 + \tfrac{1}{3} CD)$$

Calculate also the depth below the water surface at which H acts.
(b) If the reservoir depth is 50 m and the water density increases from 1000 kg/m^3 at the surface to 1100 kg/m^3 at the bottom, calculate the vertical hydrostatic force on a horizontal circular plate of diameter 10 m at the bottom of the reservoir. Calculate also the horizontal component of the hydrostatic force on a wall 10 m wide which is inclined at 60° to the horizontal.
(Answers: 40.4 MN, 126.7 MN)

5.8 (a) Figure P5.8 shows a flat plate which is immersed vertically in water to a depth D. The shape of the plate is given by

$$y = \frac{D^2 - z^2}{2D}$$

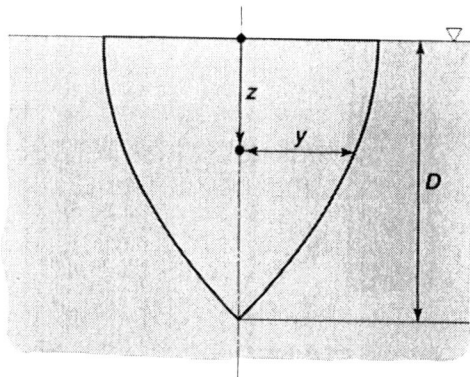

Figure P5.8

where y is the half width of the plate a vertical distance z below the surface. From first principles show the following:

(i)   The surface area of the plate $A = \frac{2}{3}D^2$
(ii)   The hydrostatic force acting on one side of the plate $H = \frac{3}{8}\rho gAD$.
(iii) The hydrostatic force acts at a distance $\frac{8}{15}D$ below the surface.

(b) A container of length 5 m and maximum width 2 m has the cross-section shown in Figure 5.8. Calculate the maximum load which the container can carry (including its own weight) without sinking. Calculate in bar the corresponding hydrostatic pressure acting at the centroid of one end of the container.
(Answers: 130.8 kN, 0.0736 bar)

5.9   (a) A rigid spherical balloon of diameter D is filled with a light gas of density $r\rho_s$ where ρ_s is the density of the atmospheric air at ground level. Show that the net upward force F experienced by the balloon at ground level is given by

$$F = \frac{1}{6}\pi D^2(1 - r)(p_L - p_H)$$

where p is the atmospheric pressure and the subscripts H and L refer to the highest and lowest points on the balloon's surface, respectively. The weight and volume of the balloon's 'skin' may be neglected.
(b) A rigid balloon of diameter 3 m is filled with helium with a density of 0.2 kg/m^3. Calculate the value of $p_L - p_H$ at the altitude where the balloon just floats without rising or falling. The acceleration due to gravity is 9.81 m/s^2. Calculate the maximum mass the balloon could lift to an altitude at which the air density is 0.45 kg/m^3.
(Answers: 5.89 Pa, 3.53 kg)

5.10   (a) Figure P5.10 shows the cross-section of a vertical rectangular barrier separating pure water of density ρ on the right-hand side from a layer of pure water of depth Z_1 on the left-hand side above a layer of silt of depth Z_2. The silt may be treated as a liquid of density ρ_2. Determine the depth Z of the water on the right-hand side if the net force on the barrier is to be zero. Is Z greater or smaller than $(Z_1 + Z_2)$, and why?

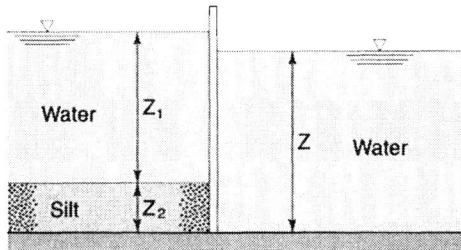

Figure P5.10

(b) If the depth of the water on the left-hand side is 1.5 m and that of the silt 0.5 m, calculate the total hydrostatic force exerted on that side of the barrier. Take the relative density of silt as 1.5 and the length of the wall as 2 m. Calculate the water depth on the right-hand side for zero net force on the barrier and the net overall moment exerted on the barrier. (Answers: 40.5 KN, 2.03 m, 1031 N.m anticlockwise)

5.11 Figure P5.11 shows the cross-section of a yacht floating in sea water of density ρ. Excluding the keel, the weight of the yacht is W and its centre of gravity a height H above the water line. The submerged section of the hull is triangular in cross-section, with apex angle 2α and height h. The weight of the keel is W_K and its centre of gravity is a depth Z_K below the water line. Show that

$$h^2 = \frac{W + W_K}{\rho g L \tan \alpha}$$

where L is the length of the yacht. The volume of the keel may be regarded as negligible. Show also that the metacentre of the yacht is at a depth

$$\tfrac{1}{3} h(1 - 2 \tan^2 \alpha)$$

below the surface. Finally show that for the yacht to be stable, the minimum weight of the keel is given by the equation

$$W_K Z_K - \frac{(W_K + W)^{3/2}(1 - 2 \tan^2 \alpha)}{3\sqrt{(\rho g L \tan \alpha)}} - WH = 0$$

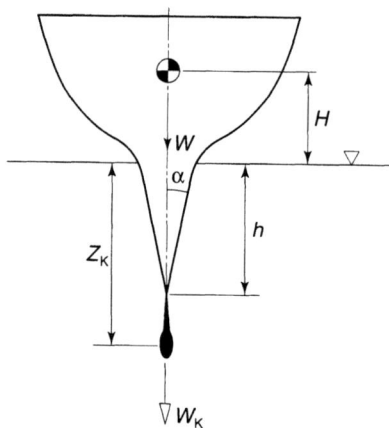

Figure P5.11

CHAPTER 6

Kinematic description of fluids in motion

We start this brief chapter by introducing the concepts of fluid particles, path-lines and streamlines, together with some of the other terms and ideas needed to describe fluid motion and apply to it the basic laws of physics. We also point out the principal simplifications which are necessary if the mathematics is not to become too demanding: with some minor exceptions, we restrict consideration to the steady flow of a constant density, single-phase fluid. We also assume that flows can be treated as one dimensional, which means that, over any cross-section through which there is flow, the fluid velocity and density (and all other fluid and flow properties) are uniform. Application of the principle of conservation of mass is shown to result in a simple but important relationship between fluid density, flow velocity and the cross-sectional area of a streamtube. The term *kinematic* in the chapter title indicates that at this stage we are concerned only with velocity variation in a flow, not with the stresses and forces associated with fluid motion.

6.1 **Fluid particles**

From section 2.3 we can see that the number of molecules contained in a cube of water of side length one-tenth of one micron (i.e. 10^{-7} m) is about thirty million while for a cube of air, at a temperature of 20°C and pressure of 1 bar, the number is about 30 000. Such a large number of molecules allows us to define average values for density, viscosity and other material properties which are independent of the volume (i.e. cube) size. In almost any situation of practical importance 0.1 μm is at least three orders of magnitude smaller than any significant dimension of a channel through which flow might occur. In normal circumstances, therefore, any changes in flow properties, such as pressure and velocity, would also be negligibly small from one side of the fluid volume to the other. Such a tiny volume of fluid, which has fluid and flow properties independent of its size, we term a *fluid particle*. A convenient way to think of a fluid particle is as a point-sized blob of fluid which has the temperature, pressure and velocity of its immediate surroundings. It should be clear that if we could mark and follow the movement of a number of fluid particles we would be able to form a visual impression of any given flow.

6.2 Steady-flow assumption

In this book we shall restrict consideration to *steady flows*, that is to flows for which the *velocity and pressure at any point in the flow do not change with time*, though in general there will be spatial variations in these quantities throughout the flowfield. Whenever a flow is created by the movement of an object, such as a car, a ship or an aircraft, through an otherwise stationary fluid, it is often possible to create a steady-flow situation by considering the flow relative to the object. This Gallilean transformation, as it is called, in which the object is brought to rest and its velocity subtracted from that of the surroundings, is restricted to objects moving at constant velocity. For example, the airflow over the wings and fuselage of an aircraft would not appear to be steady when seen by an observer on the ground, but could be regarded as steady relative to the aircraft if it were flying at constant speed at a fixed altitude.

6.3 Streamlines, streamsurfaces and streamtubes

Figure 6.1 shows the cross-section of a stationary aerofoil with fluid flowing steadily over it from left to right. Each of the lines with arrows on them represents the *pathline* followed by a fluid particle. Of the eight pathlines shown, five pass over the upper (suction) surface and three over the lower (pressure) surface of the aerofoil. In steady flow, the pathline of a particle is identical with a *streamline which is defined as a line in a flow everywhere tangent to the fluid-velocity vector at a given instant*. One consequence of this definition is that streamlines can never cross, and another is that there is no flow across a streamline so that the outline of any cross-section through a solid object must also represent a streamline. In the same way, each of the other streamlines can be thought of as

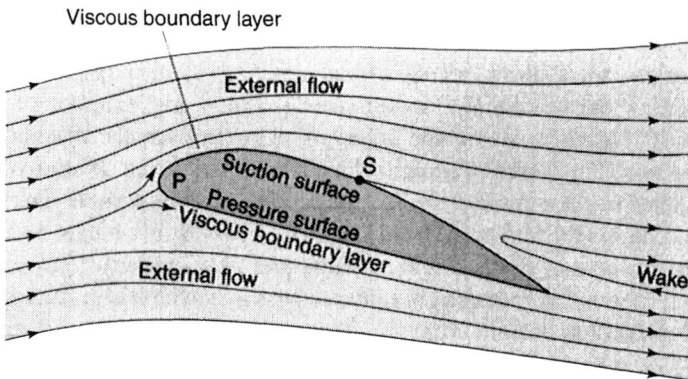

Figure 6.1 *Flow past a stationary aerofoil*

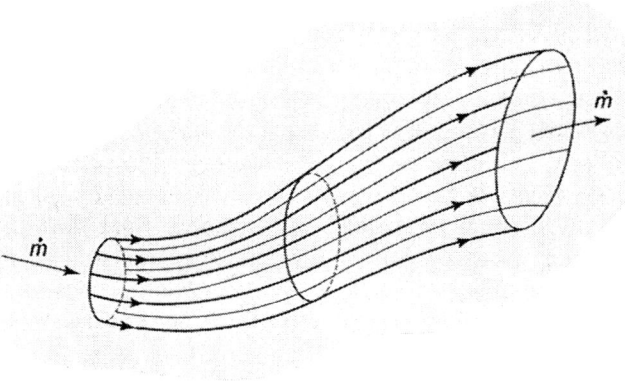

Figure 6.2 *Definition of a streamtube*

representing a section through a *streamsurface*, i.e. a *surface made up of stream-lines*. The middle streamline in Figure 6.1 is said to be a *dividing streamline* because all the streamlines above it pass over the upper surface of the aerofoil and all those below it pass over the lower surface. The dividing streamline intercepts the aerofoil at a point P on its leading edge the location of which depends on the angle of attack α between the aerofoil and the approach flow. For a solid aerofoil, the velocity at this interception point must be zero and such a point is called a *stagnation point*, the word stagnation meaning that the fluid is not moving (i.e. stagnant). For the same reason, the dividing streamline is also called a *stagnation streamline*.

As shown in Figure 6.2, if the cross-section of a streamsurface taken across the flow is a closed loop, we then have what is called a *streamtube*. Since stream-surfaces and streamtubes are made up of streamlines, there can be no flow across them and a streamtube can be thought of as representing the interior wall of a duct such as a tube or pipe.

6.4 No-slip condition and the boundary layer

Although it is obvious that fluid cannot pass through a solid surface, and so the component of velocity normal to any solid surface must be zero, it is primarily a matter of experimental observation that the *component of velocity tangential to the surface is also zero*. According to this *no-slip condition*, in the immediate vicinity of a solid surface a consequence of viscosity is that the fluid is brought to rest (or, more generally, to the same velocity as the surface so that the relative velocity is zero). In essence, the fluid sticks to the surface. The change from zero to non-zero fluid velocity takes place across a very thin layer of fluid in the immediate vicinity of the surface called the *boundary layer*. The variation of flow properties within the boundary layer results from a balance between fluid momentum, viscous stress and pressure gradient. Boundary-layer flow is characterised by particles moving along streamlines almost parallel to the surface at velocities which increase very rapidly with distance from the surface.

In the case of the aerofoil, it is found that the surface pressure on the suction surface decreases with distance from the leading edge up to about one-third chord distance and then begins to increase again. Although the absolute pressure level is normally of little importance in low-speed fluid flow problems, the development of any flow is sensitive to streamwise pressure variations. This is especially so for boundary layer flows, particularly any increase in pressure since this will oppose forward motion of the fluid, and can lead to *separation* of the boundary layer from the surface. Boundary-layer separation is shown in Figure 6.1 to occur at point S on the suction surface. A boundary layer also develops on the pressure surface but the pressure gradient is less severe and separation is unlikely. Whether or not boundary-layer separation does occur, there will be a region downstream of the aerofoil, called the *wake*, which is also low in both velocity and momentum.

Outside the boundary layer the flow can be analysed assuming the fluid is *inviscid*, i.e. it has *zero viscosity*. The concept of the boundary layer and of dividing the flow over a surface into a thin region, where viscosity is crucially important, and an outer inviscid flow was devised in 1904 by the German engineer Ludwig Prandtl (1875–1953). Even today, when immensely powerful computers are widely available to practising engineers and it might be thought that complete calculations of any flow should be possible, in fact this is not the case and Prandtl's boundary-layer theory is still widely used in the analysis of flow problems.

Although we shall make a start on the analysis of inviscid flows in the next two chapters, the detailed study of both inviscid flowfields and boundary layers is beyond the scope of this text.

6.5 Other simplifying assumptions

6.5.1 *Single-phase flow*

In Chapter 2 we saw that substances can exist in four different forms or phases (solid, liquid, vapour or gas), often depending upon the temperature and pressure to which they are subjected. Although many industrial processes involve flows in which two or more phases are present (multiphase flows), these flows are very difficult to analyse and so throughout this text we shall restrict attention to single-phase fluid flow. What this means is that we exclude from consideration the flow of:

- liquids containing bubbles of gas or vapour, as would occur in boiling and cavitation (a phenomenon we explain in section 8.10)

- immiscible liquids (i.e. liquids which do not mix but tend to separate) such as oil and water where one liquid forms droplets within the other
- liquids containing solid particles, such as blood (blood cells in plasma) or lubricating oil contaminated with metal cuttings by machining operations
- gases containing liquid droplets, for example the mixing of hot gas with atomised liquid fuel sprayed into a combustion chamber
- gases containing solid particles such as pollutants.

6.5.2 *Incompressible, isothermal and adiabatic flow*

Another major simplification we make in this book is that any of the properties of a fluid which affect its flow (i.e. density, viscosity, surface tension, and compressibility) are constant and uniform throughout the flowfield under consideration. Since all fluid properties depend to some extent on the fluid temperature, we are basically restricting consideration to constant temperature, or *isothermal*, flows. This restriction, in turn, implies both that changes in fluid pressure are relatively small and also that there is negligible transfer of thermal energy (heat) to or from the fluid.

Flows for which density changes due to pressure variations are negligible are termed *incompressible* while any process which does not involve heat transfer is said to be *adiabatic*. Because the density of any liquid is almost independent of pressure, liquid flows are invariably incompressible. For gases, which are highly compressible, the assumption of incompressibility requires that the Mach number M (defined in Chapter 3 as the ratio of flowspeed V to the speed of sound c) is less than about 0.3, i.e.

$$M = \frac{V}{c} < 0.3.$$

Values for the speed of sound are listed in Tables A.2–A.5 in the appendix.

6.5.3 *One-dimensional flow*

There are many flows in which, at certain locations, we can identify a main flow direction. The justification for this statement is evident from most of the figures in Chapter 1: the discharge from a centrifugal pump (Figure 1.2); the flow through a nozzle (Figure 1.7); a turbofan engine (Figure 1.8); a pipe bend (Figure 1.11); a rocket engine (Figure 1.12); a jet pump (Figure 1.13) and a cascade of guidevanes (Figure 1.14). The flow induced by a propeller, over a supersonic aerofoil and of water vapour emitted from a cooling tower (whether visible or invisible), are further examples illustrated in Figure 6.3.

Although the flows in most of the internal-flow examples shown in Chapter 1

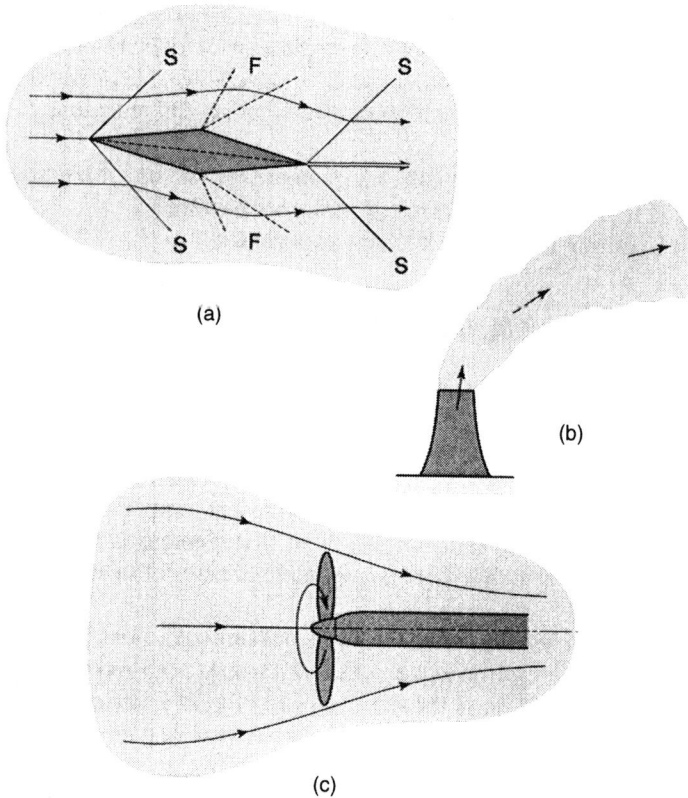

Figure 6.3 *Principal flow directions for various external flow situations: (a) double-wedge supersonic aerofoil with shock waves (S) and expansion fans (F); (b) plume from smoke stack or cooling tower; (c) propeller*

are extremely complex, we can make significant progress in their analysis by consideration of changes in the average conditions between inlet and outlet while ignoring (with some exceptions) the interior details of the flow. External flows, such as those shown in Figure 6.3, are usually more difficult to deal with and for the most part are outside the scope of this text.

From now on, we shall assume that over any cross-section of a flow the velocity V and pressure p are uniform and we account only for variations in these flow properties from location to location. The assumption that flow properties such as V and p depend only upon location s in this way, but do not vary across the cross-section, is called the *one-dimensional approximation*. The fact that the flow of real fluids is affected by viscosity and the associated no-slip condition means that the uniform-velocity assumption is certainly invalid in the immediate vicinity of any surface and may well be of limited validity in interior regions of a flow. The uniform pressure assumption is quite different in character and is usually regarded as valid wherever streamline curvature is small and totally inappropriate in situations where the streamlines are strongly curved, such as in a tornado where there is always a rapid decrease in pressure towards the eye of the tornado.

As we shall show in the next section, for steady flow of a constant-density fluid, the velocity variation $V(s)$ results directly from the shape of the streamtube through which flow occurs. In Chapter 7 we shall show that these velocity changes are accompanied by pressure variations $p(s)$ and in Chapter 9 we shall derive a form of the momentum equation which will enable us to calculate the hydrodynamic forces which a moving fluid exerts on the surfaces with which it is in contact (Chapter 10).

6.6 Continuity equation (mass-conservation equation)

The cross-sectional area A of the streamtube shown in Figure 6.4 varies in some specified way with distance along the streamtube s. As already stated, we assume that the flow through the streamtube is steady and also adopt the one-dimensional assumption that all fluid and flow properties are uniform across any given cross-section but can vary from location to location. Our aim now is to find the variation of the fluid velocity V with location s as a consequence of the area change $A(s)$.

The basis for our analysis is the principle of conservation of mass according to which fluid is neither created nor destroyed. For a steady flow, this principle requires that the same mass of fluid flows across every cross-section of the streamtube in a given time.

We start by considering an infinitesimal slice of fluid of thickness δs at some

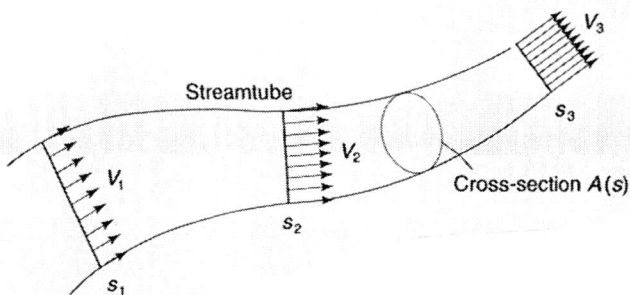

Figure 6.4 *Area and velocity variation along a streamtube*

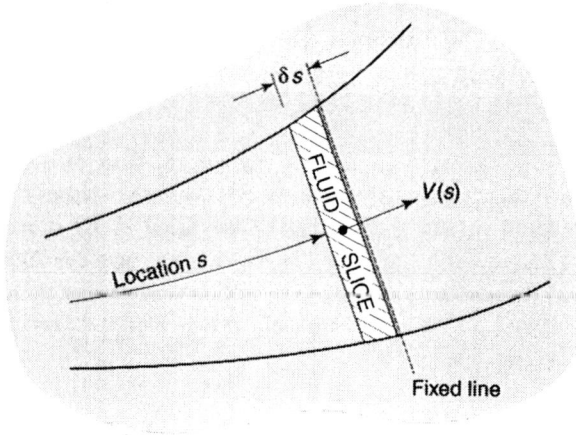

Figure 6.5 *Elemental fluid slice moving along a streamtube*

location a distance s along the streamtube, as shown in Figure 6.5. This slice has volume

$$\delta \vartheta = A \, \delta s,$$

and mass

$$\delta m = \rho \, \delta \vartheta = \rho A \, \delta s$$

where ρ is the density of the fluid within the slice (note that at this stage we do not need to assume that the density remains constant along the streamtube).

If the velocity of the fluid particles in the slice is V, then the slice will move a distance equal to its thickness δs in time δt, given by

$$\delta t = \frac{\delta s}{V}.$$

If we use this relation to substitute for δs in the expressions for $\delta \vartheta$ and δm above, we have

$$\delta \vartheta = AV \, \delta t \quad \text{and} \quad \delta m = \rho AV \, \delta t$$

or

$$\frac{\delta \vartheta}{\delta t} = AV \quad \text{and} \quad \frac{\delta m}{\delta t} = \rho AV.$$

We observe now that $\delta m/\delta t$ represents the *mass of fluid which crosses a section of the streamtube* (in this case the section which coincides with the elemental slice) *per unit time*. This quantity is constant for a steady flow and is called the *mass flow rate* $\dot{m}$, i.e.

$$\boxed{\dot{m} = \rho A V = \text{constant.}} \qquad (6.1)$$

Equation (6.1) is referred to as the *continuity equation*. The historical origin of this name is unclear with some fluid dynamicists suggesting it reflects the continuum character of density and velocity while others feel it denotes the continuous nature of fluid flow. Equation (6.1) is also called the *mass-conservation equation*.

The quantity $\delta\vartheta/\delta t$ represents the *volume of fluid which crosses a section of the streamtube per unit time and is termed the volumetric (or volume) flowrate* $\dot{Q}$, i.e.

$$\boxed{\dot{Q} = AV.} \qquad (6.2)$$

Equations (6.1) and (6.2) are two extremely important results, one or other of which is used in every one-dimensional, steady-flow analysis, including situations where the density changes as a consequence of heating, cooling or pressure changes.

Example 6.1
Water flows through a circular nozzle which contracts from an inlet area of 0.05 m^2 to an outlet of area 0.01 m^2. If the mass flowrate is 110 kg/s, calculate the volumetric flowrate and the water velocity at inlet and outlet.

Solution

$$\dot{m} = 110 \text{ kg/s}; \quad A_1 = 0.05 \text{ m}^2; \quad A_2 = 0.01 \text{ m}^2; \quad \rho = 1000 \text{ kg/m}^3$$

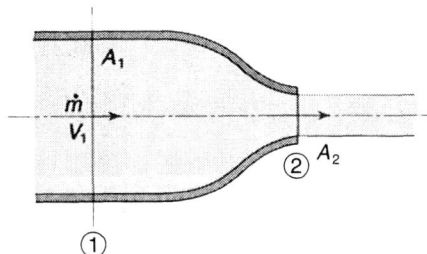

Figure E6.1

From the mass-conservation equation, we have $\dot{m} = \rho\dot{Q}$ so that

$$\dot{Q} = \frac{\dot{m}}{\rho} = \frac{110}{1000} = 0.11 \text{ m}^3/\text{s}$$

From $\dot{Q} = AV$, we have $\dot{Q} = A_1 V_1 = A_2 V_2$ so that

$$V_1 = \frac{\dot{Q}}{A_1} = \frac{0.11}{0.05} = 2.2 \text{ m/s}$$

and

$$V_2 = \frac{\dot{Q}}{A_2} = \frac{0.11}{0.01} = 11 \text{ m/s}.$$

Example 6.2

A cryogenic wind tunnel is being designed to develop the aerodynamic performance of a Formula One racing car. The wind tunnel will operate with an air density of 8 kg/m³ at an airspeed in the working section of 90 m/s. The working section is to have a cross-section 1 m high and 2 m wide and will be just downstream of a contraction from a plenum chamber with a cross-sectional area of 10 m². Calculate the air mass flowrate through the wind tunnel and the airspeed in the plenum chamber.

Solution

$$V_2 = 90 \text{ m/s}; \quad \rho = 8 \text{ kg/m}^3; \quad A_1 = 10 \text{ m}^2; \quad A_2 = 2 \text{ m}^2$$

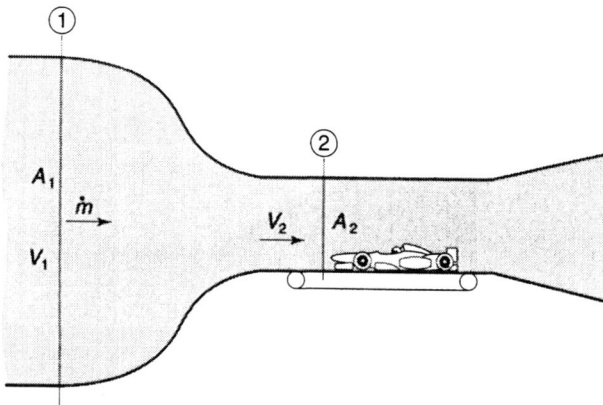

Figure E6.2

From the mass-conservation equation, $\dot{m} = \rho A V$, we have

$$\dot{m} = 8 \times 2 \times 90 = 1440 \text{ kg/s}$$

Also, $\dot{Q} = A_1 V_1 = A_2 V_2$, so that

$$V_1 = \frac{A_2 V_2}{A_1} = \frac{2 \times 90}{10} = 18 \text{ m/s}.$$

Example 6.3
Helium flows through a nozzle with an outlet diameter of 50 mm. Calculate the maximum mass flowrate for which the flow can be considered as incompressible. The density of helium can be taken as 0.17 kg/m^3 and the speed of sound in helium as 1007 m/s.

Solution

$$D = 0.05 \text{ m}; \quad \rho = 0.17 \text{ kg/m}^3; \quad c = 1007 \text{ m/s}$$

For a flow to be considered incompressible we require

$$M < 0.3.$$

Since $M = V/c$ we require

$$V < 0.3 \times 1007 = 302.1 \text{ m/s}.$$

We have

$$A = \tfrac{1}{4} \pi D^2 = 1.99 \times 10^{-4} \text{ m}^2 \quad \text{and} \quad \dot{m} = \rho A V,$$

so that

$$\dot{m} < 0.17 \times 1.99 \times 10^{-4} \times 302.1 = 0.010 \text{ kg/s}.$$

6.7 Average velocity

In a real flow through a duct, the fluid velocity varies from zero at the duct walls (the no-slip condition) to a maximum usually somewhere close to the duct axis. No matter how complex the velocity variation (also called the velocity distribution or profile), it is often convenient to define an *average fluid velocity* $\bar{V}$ using

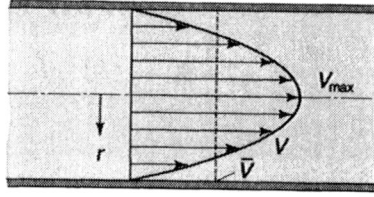

Figure 6.6 *Viscous Liquid flow through a long circular pipe*

equation (6.1) or (6.2) i.e.

$$\bar{V} \equiv \frac{\dot{m}}{\rho A} \equiv \frac{\dot{Q}}{A} \qquad (6.3)$$

The quantity $\bar{V}$ is also referred to as the *mean velocity* or *bulk mean velocity*.

It is convenient and appropriate when applying the one-dimensional approximation to identify the velocity V with the average velocity $\bar{V}$, but it has to be appreciated that other quantities involving $\bar{V}$, such as the kinetic-energy flowrate $\frac{1}{2}\dot{m}\bar{V}^2$, do not accurately represent the average values of these quantities for a real flow. For example, for the flow of a viscous liquid through a long circular pipe at low flowrates, the velocity variation across the pipe is parabolic, as depicted in Figure 6.6, and we find $\bar{V} = \frac{1}{2} V_{\max}$, $V_{\max}$ being the centreline velocity. The true kinetic-energy flowrate for this flow is given by $\dot{m}\bar{V}^2$, i.e. double the value for a flow with uniform velocity $\bar{V}$. Fortunately the velocity distribution for many flows of engineering importance is much flatter (i.e. closer to uniform velocity) than the parabolic profile and the one-dimensional approximation usually leads to results of acceptable accuracy.

6.8 Flow of a fluid of constant density

Equations (6.1) and (6.2) are related as follows

$$\dot{m} = \rho A V = \rho \dot{Q} = \text{constant.}$$

If the fluid density ρ is constant, as will be the case for all problems considered in this text, then we have

$$\dot{Q} = AV = \text{constant} \qquad (6.4)$$

which shows that if the area of the streamtube $A(s)$ decreases with distance along the streamtube s, the fluid velocity $V(s)$ must increase; if A increases, V

must decrease. Leonardo da Vinci (1452–1519) was certainly aware of the continuity equation in this form and there is some evidence that it was known 1400 years earlier to Sextus Julius Frontinus (40–103 A.D), a Roman military engineer.

As we shall show in the next chapter, increases in the velocity of a fluid are accompanied by decreases in its pressure. For the flow of a liquid, if the reduction in pressure is sufficiently great, bubbles of dissolved gas or vapour may appear in the flow, a phenomenon known as *cavitation*. Although the presence of gas or vapour bubbles would violate our assumption of single-phase flow and render our analysis invalid (as we shall show in section 8.10), the analysis is of great value in predicting the onset of cavitation.

For the flow of a gas, in the absence of heat transfer, a decrease in pressure is inevitably accompanied by a decrease in density. Provided that the Mach number does not exceed a value of about 0.3, the density decrease is usually negligible (e.g. for air it is less than 5%) and the flow may still be regarded as incompressible. For higher Mach numbers, compressibility effects, such as choking and shock waves, become important but their treatment is also beyond the scope of this text.

6.9 Summary

In this chapter we have introduced some of the terminology and simplifications which enable us to begin to describe and analyse practical fluid-flow problems. The principle of conservation of mass applied to steady one-dimensional flow through a streamtube of varying cross-sectional area resulted in the continuity equation which relates mass flowrate $\dot{m}$, volumetric flowrate $\dot{Q}$, fluid velocity V, fluid density ρ, and cross-sectional area A:

$$\boxed{\dot{m} = \rho\dot{Q} = \rho A V = \text{constant.}}$$

For a constant-density fluid this result shows that fluid velocity increases if the cross-sectional area decreases, and vice versa.

The student should be able to:

- explain what is meant by the following terms:
  fluid particle   steady flow   streamline   streamsurface   streamtube
  no-slip condition   boundary layer   single-phase flow   incompressible
  isothermal   adiabatic   one-dimensional flow   average velocity
- apply the continuity equation $\dot{m} = \rho A V$ to duct flow and state the assumptions made in its derivation
- apply the equation for the volumetric flowrate $\dot{Q} = A V$ to duct flow and state the assumptions made in its derivation

6.10 Self-assessment problems

6.1   Liquid medication of density 990 kg/m³ is injected from a hypodermic
syringe with an internal diameter of 10 mm through a needle with an
internal diameter of 0.3 mm. If it takes 30 s to inject 2 ml of liquid,
calculate the mass flowrate and the liquid velocities within the syringe
and the needle. Assume the plunger moves at constant speed.
(Answers: 6.6×10^{-5} kg/s, 8.5×10^{-4} m/s, 0.94 m/s)

6.2   Two pipes, one of internal diameter 0.5 m and the other 1 m are
connected as shown in Figure P6.2 to a pipe of internal diameter 1.2 m.
Oil with a density of 880 kg/m³ flows through the pipe system at a total
rate of 15 000 tonne/hr. If the liquid velocity in each of the two smaller
pipes is the same, calculate this velocity and the corresponding volumetric
and mass flowrates, and also the velocity in the large outlet pipe.
(Answers: 4.82 m/s, 0.95 m³/s, 3.79 m³/s, 3000 t/hr, 12000 t/hr,
4.19 m/s)

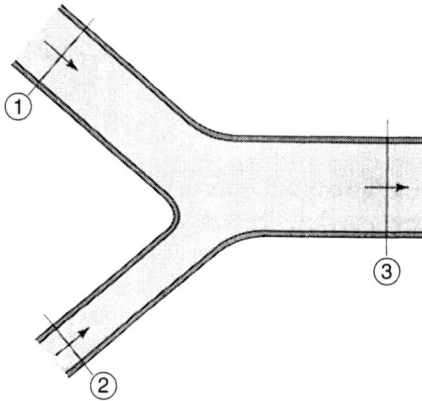

Figure P6.2

6.3   Hot gas with a density of 0.4 kg/m³ is exhausted from a rocket engine
through a nozzle of exit diameter 1 m. If the mass flowrate through the
nozzle is 370 kg/s, calculate the exhaust gas velocity and the volumetric
flowrate. If the soundspeed of the gas is 550 m/s, calculate the Mach
number of the exhaust gas flow. Can the exhaust flow be considered as
incompressible?
(Answers: 1178 m/s, 925 m³/s, 2.14, no)

6.4   A water jet 50 mm in diameter impinges on a cone as shown in Figure
P6.4. If the water velocity has the same magnitude at all points in the
flow, calculate the thickness of the liquid layer at a location where the

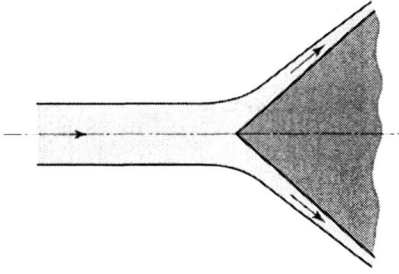

Figure P6.4

cone diameter is 0.5 m. If the mass flowrate of the jet is 16 kg/s, calculate the flowspeed.
(Answers: 1.2 mm, 8.15 m/s)

CHAPTER 7

Bernoulli's equation

In this chapter we apply Newton's second law of motion to derive, first, Euler's equation, which is a differential equation connecting the pressure, velocity and height of a fluid particle moving steadily along a streamline in an inviscid fluid. By integrating Euler's equation for a fluid of constant density, we obtain Bernoulli's equation which, in spite of the underlying restrictions, is one of the most important and practically useful equations of fluid mechanics. It is shown that the terms in Bernoulli's equation can be interpreted either as pressures, as forms of energy or as the heights of fluid columns.

7.1 The net force on an elemental slice of fluid flowing through a streamtube

In the previous chapter we analysed the motion of a slice of fluid of elemental thickness δs as it moved along a streamtube. For steady flow we found that the principle of conservation of mass leads to the continuity equation

$$\dot{m} = \rho \dot{Q} = \rho A V$$

where $\dot{m}$ is the mass flowrate through the streamtube, $\dot{Q}$ the volumetric flowrate, ρ the fluid density, V the magnitude of the fluid velocity and A the cross-sectional area of the streamtube. To go further we need to introduce another of the basic laws of classical physics, the principle of conservation of momentum which is usually referred to as Newton's second law of motion and written

$$F = ma$$

where the net force F acting on a fixed mass m results in an acceleration of the mass of magnitude a in the direction of F. It is essential to understand that F and a are vector quantities and so have not only magnitude but also direction.

  As shown in Figure 7.1, forces acting on the slice of fluid arise from its weight and the pressures acting on each of its three faces. Since we are treating the fluid as inviscid (i.e. as having zero viscosity), any forces due to viscous shear stresses are identically equal to zero. This '*ideal fluid*' assumption is clearly rather drastic

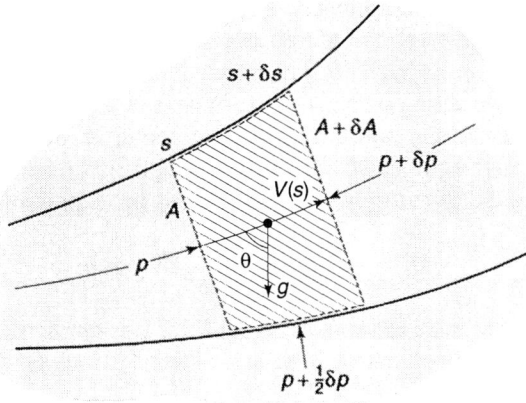

Figure 7.1 *Elemental fluid slice moving along a streamtube*

since in Chapter 2 we identified viscosity as the essential property of any fluid. Nevertheless it is a simplification which allows us to make progress in the analysis of fluid motion and, as it turns out, produces remarkably accurate results for a wide range of practical problems.

The net force δF in the s-direction which acts on the slice is made up of four components, as follows:

- pressure force pA
- pressure force $-(p + \delta p)(A + \delta A)$
- pressure force $(p + \frac{1}{2} \delta p) \delta A$
- the component of the weight of the slice in the s-direction $-\delta W \cos \theta$

so that

$$\delta F = pA - (p + \delta p)(A + \delta A) + (p + \tfrac{1}{2} \delta p) \delta A - \delta W \cos \theta$$

In writing these terms we have assumed that the pressure p is uniform across any cross-section of the streamtube (i.e. the one-dimensional assumption) but varies with distance s along it. The terms are taken as positive in the direction of increasing s and negative in the opposite direction. The cross-sectional area of the streamtube A has also been assumed to vary with s. Although Figure 7.1 shows A increasing with s, this in no way restricts the analysis which is valid whether A increases or decreases. The angle between the velocity vector at any location along the streamtube and the vertical is denoted by θ.

The third term on the right-hand side of the equation for δF represents the force due to the pressure acting on the small section of the surface of the streamtube which coincides instantaneously with the moving slice, i.e. a strip of the surface rather like a piece cut from an inner tube. The average pressure acting on this strip must have a value somewhere between p and $p + \delta p$ and has been taken as the simple average $p + \frac{1}{2} \delta p$ though, as we shall see shortly, the factor $\frac{1}{2}$ is not important. Just as for the horizontal component of the elemental force due to the hydrostatic pressure acting on a curved surface (see section 5.7), the net force in the s-direction due to the pressure $(p + \frac{1}{2} \delta p)$ acting on the strip of streamtube surface is $(p + \frac{1}{2} \delta p) \times$ projected area, where the projected area is δA.

If we now expand and simplify by cancellation the expression for δF, we have

$$\delta F = -\delta p (A + \tfrac{1}{2} \delta A) - \delta W \cos \theta.$$

Since our elemental slice is infinitesimally thin, it is permissible to neglect the area change δA in comparison with the area A itself (which is why the factor $\frac{1}{2}$ is not important), so that

$$\delta F = -\delta p \, A - \delta W \cos \theta$$

The weight δW of the elemental slice is given by

$$\delta W = \delta m \, g = \rho \, \delta \vartheta = \rho \, A \, \delta s \, g$$

where δm is the mass of the slice and $\delta \vartheta$ its volume, and g is the acceleration due to gravity.

If we substitute for δW in the expression for δF, we have

$$\delta F = -\delta p \, A - \rho A \, \delta s \, g \cos \theta$$

or

$$\delta F = -\delta p \, A - \rho A \, \delta z' \, g \tag{7.1}$$

wherein we have made use of the fact that the vertical height change $\delta z'$ corresponding to the distance δs is given by $\delta z' = \delta s \cos \theta$ (see Figure 7.1). Just as in Chapter 4, we use the symbol z' to denote altitude and z to denote depth so that $\delta z = -\delta z'$.

7.2 Acceleration of fluid slice

Since we have restricted our attention to steady flow, it may seem to be a contradiction that we are now discussing acceleration of the fluid. However, as stated in section 6.2, where this restriction was introduced, steady flow of a fluid implies that the fluid velocity at any fixed point in a flowfield is always the same but can vary from point to point. What this means is that the velocity V of each fluid particle can change as it moves through the flowfield and so the particle experiences acceleration.

If the acceleration of a particle with instantaneous velocity V at time t is a, by

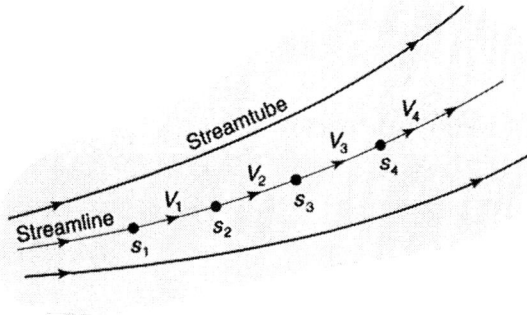

Figure 7.2 *Velocity variations for steady flow along a streamline*

definition we have

$$a = \frac{dV}{dt} = \frac{dV}{ds}\frac{ds}{dt}$$

wherein we have made use of the rule of differential calculus for differentiation of a function of a function: in this case V is a function of s which itself is a function of t. We see why the latter step is useful when we realise that, also by definition, the particle velocity V in the direction s is

$$V = \frac{ds}{dt},$$

so the acceleration of a fluid particle in the s-direction can be written as

$$a = V\frac{dV}{ds}$$

EXAMPLE 7.1

A liquid flows at a constant volumetric flowrate $\dot{Q}$ through a duct which decreases in area A such that $A/A_0 = s_0/s$, where s is the distance along the duct and A_0 is the area at $s = s_0$. Derive an expression for the fluid acceleration. If $s_0 = 1$ m, $A_0 = 1$ m^2 and $\dot{Q} = 0.2$ m^3/s, calculate the fluid velocity and acceleration at $s = 5$ m.

SOLUTION

We have $\dot{Q} = AV$ so that

$$V = \frac{\dot{Q}}{A} = \frac{\dot{Q}s}{A_0 s_0}$$

from which

$$\frac{dV}{ds} = \frac{\dot{Q}}{A_0 s_0}$$

Then from $a = V(dV/ds)$, we have

$$a = \left(\frac{\dot{Q}}{A_0 s_0}\right)^2 s.$$

For the numerical part of the problem we have

$$s_0 = 1 \text{ m}; \quad A_0 = 1 \text{ m}^2; \quad \dot{Q} = 0.2 \text{ m}^3/\text{s}; \quad s = 5 \text{ m}$$

At $s = 5$ m:

$$A = \frac{A_0 s_0}{s} = \frac{0.1 \times 1}{5} = 0.02 \text{ m}^2$$

therefore

$$V = \frac{\dot{Q}}{A} = \frac{0.2}{0.02} = 10 \text{ m/s}$$

and

$$a = \left(\frac{\dot{Q}}{A_0 s_0}\right)^2 s = \left(\frac{0.2}{0.1 \times 1}\right)^2 \times 5 = 20 \text{ m/s}^2.$$

7.3 Euler's equation

We are now in a position to apply Newton's second law of motion to our fluid slice since we have expressions for the net force acting on the slice in the s-direction and for the acceleration of the fluid particles which make up the slice. We start with the basic form of Newton's second law

$$\delta F = \delta m \, a$$

From section 7.1 we have the mass of the slice $\delta m = \rho A \, \delta s$ and from section 7.2 the acceleration $a = V(dV/ds)$, so that

$$\delta F = \rho A \, \delta s \, V \frac{dV}{ds} \tag{7.2}$$

By equating the two expressions for δF (equations (7.1) and (7.2)) and cancelling out the area A, we have

$$-\delta p - \rho \; \delta z' \; g = \rho \; \delta s \; V \frac{dV}{ds}.$$

We now divide through by δs and rearrange to find

$$\frac{\delta p}{\delta s} + \rho g \frac{\delta z'}{\delta s} + \rho V \frac{dV}{ds} = 0$$

which, in the limit $\delta s \rightarrow 0$, gives

$$\frac{dp}{ds} + \rho g \frac{dz'}{ds} + \rho V \frac{dV}{ds} = 0. \tag{7.3}$$

Equation (7.3) is a first-order ordinary differential equation which connects the pressure, velocity and density for the steady flow of an inviscid fluid and is known as *Euler's equation* (see section 3.12).[*]

As was the case for the mass-conservation equation (6.1), Euler's equation is valid whether or not the density is constant. We note too that equation (7.3) is independent of the area A and is, in fact, valid along a streamline for any steady inviscid flow. This general validity of Euler's equation means, for example, that it applies to both internal and external flows. It should be evident that for a fluid at rest (i.e. $V = 0$), equation (7.3) reduces to the hydrostatic equation

$$-\frac{dp}{dz'} = \frac{dp}{dz} = \rho g. \tag{4.4}$$

Unless the density ρ is taken as constant, the solution (i.e. integration) of Euler's equation is difficult and beyond the scope of this text.

7.4 **Bernoulli's equation**

If the density is taken as constant, then equation (7.3) is easily integrated and we have

$$\boxed{p + \rho g z' + \tfrac{1}{2} \rho V^2 = \text{constant}} \tag{7.4}$$

which is known as *Bernoulli's equation* (or theorem) after the Swiss mathematician Daniel Bernoulli (1700–82) who included a form of it in a textbook he

---

[*] To be precise, equation (7.3) is a form of a much more general set of equations derived by Euler for the flow of an inviscid fluid.

published in 1738. The title of the textbook, *Hydrodynamica*, introduced the name hydrodynamics for the study of fluids in motion. In fact, it was not until 1755 that Bernoulli's close friend Leonhard Euler gave a complete derivation of what we now call Bernoulli's equation. The constant of integration in equation (7.4) is called the *Bernoulli constant*.

Since we shall make extensive use of Bernoulli's equation throughout the rest of this text, it is important to be aware of the key assumptions on which its validity depends:

- steady flow
- constant-density fluid
- inviscid fluid.

The first two of these assumptions were discussed in Chapter 6 while the third was introduced in section 7.1 above. The assumption that the fluid is inviscid means that the flow is frictionless and no mechanical energy is converted to heat. The term 'loss free' is often used for such flows.

As was the case for Euler's equation, subject to the additional restriction of constant density, Bernoulli's equation is valid along any streamline for the steady flow of an inviscid fluid. Although the Bernoulli constant on the right-hand side of equation (7.4) can in principle vary from streamline to streamline, we shall restrict consideration to one-dimensional flows where the Bernoulli constant is uniform throughout the flowfield.

EXAMPLE 7.2

Calculate the Bernoulli constant for water flowing through a pipe at zero altitude at a speed of 10 m/s if the water pressure is 1 bar. If the elevation of the pipe falls by 20 m and the flowspeed decreases to 2 m/s, what is the new fluid pressure?

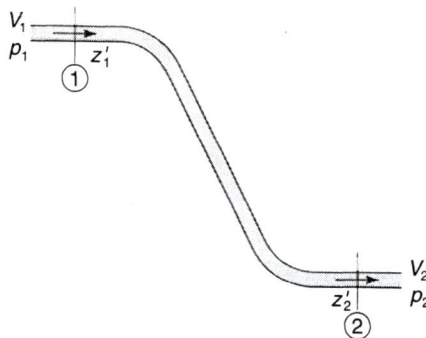

Figure E7.2

SOLUTION

$z'_1 = 0$; $V_1 = 10$ m/s; $p_1 = 10^5$ Pa; $z'_2 = -20$ m; $V_2 = 2$ m/s; $\rho = 10^3$ kg/m^3; $g = 9.81$ m/s^2

The terms in Bernoulli's equation are

- $p_1 = 10^5$ Pa
- $\rho g z'_1 = 0$
- $\frac{1}{2} \rho V_1^2 = \frac{1}{2} \times 10^3 \times 10^2 = 5 \times 10^4$ Pa
- the Bernoulli constant $= 10^5 + 0 + (5 \times 10^4) = 1.5 \times 10^5$ Pa or 1.5 bar.

We have $z'_2 = -20$ m, $V_2 = 2$ m/s, so that

$$\rho g z'_2 = -10^3 \times 9.81 \times 20 = -1.96 \times 10^5 \text{ Pa}$$

and

$$\frac{1}{2} \rho V_2^2 = 0.5 \times 10^3 \times 2^2 = 2 \times 10^3 \text{ Pa}.$$

If the flow is steady, of constant density and frictionless, the Bernoulli constant remains unchanged so that

$$p_2 - 1.96 \times 10^5 + 2 \times 10^3 = 1.5 \times 10^5$$

and

$$p_2 = 3.44 \times 10^5 \text{ Pa} \quad \text{or} \quad 3.44 \text{ bar}.$$

Note that it was important to take z'_2 as negative because the elevation of the pipe had fallen.

7.5 Interpretations of Bernoulli's equation

7.5.1 *Pressure*

The first term of Bernoulli's equation is the pressure which would be sensed by an observer moving with the fluid and is called the *static* or *thermodynamic pressure*. From the principle of dimensional homogeneity, which we discussed in section 3.5, if one term in equation (7.4) is a pressure then each of the other terms, including the constant of integration, must have the dimensions of pressure. In fact, it is common practice to regard each of these terms individually, and also certain combinations of them, as pressures:

$$\rho g z' = -\rho g z = -\text{hydrostatic pressure}$$
$$p + \rho g z' = \text{piezometric pressure}$$
$$\tfrac{1}{2}\rho V^2 = \text{ dynamic pressure}$$
$$p + \tfrac{1}{2}\rho V^2 = \text{stagnation pressure} = p_0$$
$$p + \rho g z' + \tfrac{1}{2}\rho V^2 = \text{total pressure} = p_T.$$

The names *hydrostatic* and *piezometric* were introduced in Chapter 4 (section 4.2), where we considered fluids at rest, while the *dynamic pressure* $\tfrac{1}{2}\rho V^2$ is the new term which arises when the fluid is in motion. The *total pressure* p_T is constant along any streamline in a steady inviscid flow of a constant-density fluid and is identically *equal to the Bernoulli constant*. For flow along a stagnation streamline, where the flow is brought to rest at a stagnation point (see Figure 6.1), the pressure at the *stagnation point* is equal to the *total pressure*. The stick man on the left in Figure 7.3 moving along the streamline at the same velocity as the fluid in his immediate vicinity would sense the local static pressure p, whereas the stationary stick man on the right would sense the total pressure p_T.

For flows where changes in the hydrostatic pressure $-\rho g z'$ along a streamline

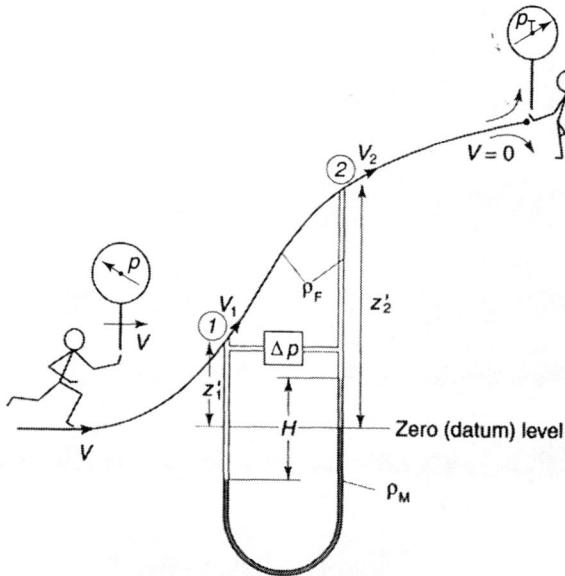

Figure 7.3 *Static and stagnation pressures*

are negligible compared to changes in the static and dynamic pressures, the *stagnation pressure* p_0 is essentially constant along a streamline. This condition applies if the streamline lies in a horizontal plane and is an excellent approximation for almost all incompressible gas flows.

EXAMPLE 7.3

The difference between the stagnation pressure and the static pressure of air of density 4.4 kg/m^3 and soundspeed 323 m/s is found to be 172 kPa. Calculate the gas velocity and determine whether the flow can be considered incompressible.

SOLUTION

$\rho = 4.4 \text{ kg/m}^3$; $c = 323 \text{ m/s}$; $p_0 - p = 172\,000 \text{ Pa}$

In this case the hydrostatic term $-\rho g z'$ is of no consequence and Bernoulli's equation reduces to

$$p + \tfrac{1}{2}\rho V^2 = p_0$$

from which

$$p_0 - p = \tfrac{1}{2}\rho V^2$$

so that

$$V = \sqrt{\frac{2(p_0 - p)}{\rho}}$$

$$= \sqrt{\frac{2 \times 172\,000}{4.4}} = 279.6 \text{ m/s}.$$

The Mach number $Ma = V/c = 279.6/323 = 0.87$, a value for which compressibility effects would not be negligible and so the use of Bernoulli's equation to determine the gas velocity would lead to error. It can be shown that the error is about 8% and it would depend upon the situation as to whether or not this was acceptable.

The dynamic pressure $\tfrac{1}{2}\rho V^2$ can be thought of as a pressure which characterises the motion of a fluid, and its value at a particular location, usually upstream, such as in the uniform flow approaching the aerofoil of Figure 6.1, is frequently chosen to make other pressures, pressure differences, or the surface shear stress τ_s non-dimensional. Several such non-dimensional quantities were introduced in section 3.12:

$$\frac{(p_{REF} - p_V)}{\frac{1}{2} \rho V^2} = \text{cavitation number}$$

where p_{REF} is a reference pressure and p_V is the saturated vapour pressure for a flowing liquid (see sections 2.12 and 8.10)

$$c_F = \frac{\tau_s}{\frac{1}{2} \rho V^2} = \text{friction factor or friction coefficient.}$$

The dynamic pressure is also used together with an appropriate area to non-dimensionalise forces such as the drag and lift forces D and L exerted on an object by the fluid flowing past it:

$$C_D = \frac{D}{\frac{1}{2} \rho V^2 A} = \text{drag coefficient}$$

$$C_L = \frac{L}{\frac{1}{2} \rho V^2 A} = \text{lift coefficient}$$

Either the projected frontal area (i.e. the area corresponding to a silhouette) or (for a wing) the planform area are frequently chosen for A.

To the above we can add

$$C_P = \frac{\Delta p}{\frac{1}{2} \rho V^2} = \text{pressure coefficient}$$

Δp is a pressure loss or pressure difference with respect to a reference pressure such as the static pressure at the same location as that for V (often equal to the barometric pressure B).

The inclusion of the factor $\frac{1}{2}$ in these definitions is conventional, and a consequence of its natural occurrence in Bernoulli's equation. It is not essential (without it all quantities are still non-dimensional) and is sometimes omitted, as is the case for the Euler number

$$Eu = \frac{p - p_{REF}}{\rho V^2}$$

7.5.2 *Energy*

The dynamic pressure, $\frac{1}{2} \rho V^2$, represents the kinetic energy of the flow per unit volume of fluid. We can see that this is so by considering a mass m of volume ϑ

moving at speed V. The kinetic energy of the mass is $\frac{1}{2} mV^2$ and its kinetic energy per unit volume therefore is $mV^2/2\vartheta$ or $\frac{1}{2} \rho V^2$ since the density $\rho \equiv m/\vartheta$. Again on the basis of the principle of dimensional homogeneity, it must be the case that each of the other terms in Bernoulli's equation can also be regarded as representing a form of energy:

p = pressure energy per unit volume
$\rho g z'$ = potential energy per unit volume
p_T = total energy per unit volume

and it follows that Bernoulli's equation itself can be thought of as an equation for the conservation of mechanical energy. In fact, provided we are consistent in our assumptions, Bernoulli's equation can be derived directly from the *First Law of Thermodynamics* which is the basis for a general energy-conservation equation. The crucial assumptions, introduced in section 6.5.2 are that there is negligible thermal energy transfer and work input to the flow.

7.5.3 *Head*

If we divide through Bernoulli's equation by ρg we find

$$\frac{p}{\rho g} + z' + \frac{V^2}{2g} = \frac{p_T}{\rho g}. \tag{7.5}$$

Once again the dimensional homogeneity argument leads to the conclusion that since z' represents altitude or height, in this form each term in Bernoulli's equation corresponds to a height or, as it is usually called, a *head*:

$$\frac{p}{\rho g} = \text{static head}$$

$$\frac{V^2}{2g} = \text{dynamic head}$$

$$\frac{p_T}{\rho g} = \text{total head}.$$

The head in each case corresponds to the vertical height of a column of fluid with the same density ρ as that of the flowing fluid.

7.6 **Pressure loss versus pressure difference**

It is important to understand the distinction between pressure difference and pressure loss. At points ① and ② on the streamline shown in Figure 7.3 the pressures, velocities and heights are connected by Bernoulli's equation as follows:

$$p_1 + \rho g z_1' + \tfrac{1}{2}\rho V_1^2 = p_2 + \rho g z_2' + \tfrac{1}{2}\rho V_2^2 = p_T \qquad (7.6)$$

or

$$(p_1 + \rho g z_1') - (p_2 + \rho g z_2') = \tfrac{1}{2}\rho(V_2^2 - V_1^2) \qquad (7.7)$$

Equation (7.7) shows that a change in velocity between points ① and ② results in a change in the piezometric pressure $(p + \rho g z')$ and this is precisely the pressure difference which would be measured by a manometer or pressure transducer as illustrated in the figure. If the velocities at points ① and ② were the same, both the manometer and the pressure transducer would indicate zero because the only change in pressure would be the hydrostatic pressure difference due to the height difference $z_2' - z_1'$ whereas the difference in the piezometric pressures is zero.

The difference in pressure associated with a velocity change becomes clearer if the hydrostatic pressure difference $\rho g(z_2' - z_1')$ is negligible since we then have

$$p_1 - p_2 = \tfrac{1}{2}\rho(V_2^2 - V_1^2). \qquad (7.8)$$

From equation (7.8) we see that an increase in velocity results in a decrease in pressure and vice versa. If we couple this statement with the constant-density form of the continuity equation (6.2), then for one-dimensional, steady flow of an inviscid, constant-density fluid through a streamtube, we can conclude that:

- if $A_2 < A_1$ then $V_2 > V_1$ and $p_2 < p_1$
- if $A_2 > A_1$ then $V_2 < V_1$ and $p_2 > p_1$.

It should be clear that whether the static and piezometric pressures increase or decrease, according to our assumptions the total pressure p_T will remain constant. In practice the effect of fluid friction at a surface, due to viscosity, is for the total pressure to decrease in the absence of thermal-energy or work input to the fluid. According to the energy interpretation of Bernoulli's equation (section 7.5.2), such a reduction in total pressure corresponds to a loss of mechanical energy. A more detailed analysis reveals that, for a viscous fluid, mechanical energy is dissipated resulting in an increase in its internal energy and hence an increase in the fluid temperature. This frictional heating is usually negligible but can become a major factor at very high gas velocities – for example, as encountered in supersonic flight or re-entry of spacecraft into the earth's atmosphere.

7.7 **Summary**

In this chapter we used Newton's second law of motion to derive Euler's equation for the flow of an inviscid fluid along a streamline. For a fluid of constant density ρ Euler's equation can be integrated to yield Bernoulli's equation

$$p + \rho g z' + \tfrac{1}{2} \rho V^2 = p_T$$

which shows that the sum of the static pressure p, −hydrostatic pressure $-\rho g z'$, and dynamic pressure $\tfrac{1}{2} \rho V^2$ is equal to the Bernoulli constant p_T. Each of the three terms on the left-hand side can be regarded as representing different forms of mechanical energy and also equivalent to the hydrostatic pressure due to a vertical column of fluid. The dynamic pressure can be thought of as measuring the intensity or strength of a flow and is frequently combined with other fluid and flow properties to produce non-dimensional numbers which characterise various aspects of fluid motion.

The student should be able to

- state Bernoulli's equation in the forms

$$p + \rho g z' + \tfrac{1}{2} \rho V^2 = \text{constant} = p_T$$

and

$$p_1 + \rho g z'_1 + \tfrac{1}{2} \rho V_1^2 = p_2 + \rho g z'_2 + \tfrac{1}{2} \rho V_2^2$$

- state the assumptions made in the derivation of Bernoulli's equation and the limitations on its applicability, i.e. to a streamline in the steady flow of an inviscid, constant-density fluid
- define the terms

  - Bernoulli constant
  - dynamic pressure and dynamic head
  - total pressure and total head
  - stagnation pressure

  in addition to the relevant terms introduced in Chapter 3

  - hydrostatic pressure and hydrostatic head
  - piezometric pressure and piezometric head

- interpret Bernoulli's equation in terms of pressure, mechanical energy, and head
- distinguish between pressure difference or change and pressure loss.

7.8 **Self-assessment problems**

7.1   Water from a reservoir flows through a pipe 2 m in diameter to the nozzles of a Pelton turbine. The vertical height between the reservoir and the turbine is 400 m. Assuming steady, one-dimensional, frictionless flow, calculate the flow velocity in the pipe and at the nozzle outlet if the nozzle diameter is 200 mm. The static pressure at outlet is the same as that at the surface.

(Answers: 0.0886 m/s, 88.6 m/s)

7.2   The mass flowrate of methane gas through a pipeline of diameter 0.5 m is 10 kg/s. Given that the density of methane is 0.66 kg/m^3, calculate the gas velocity. If the pipeline contracts linearly to a diameter of 0.35 m over a distance of 0.5 m, calculate the new gas velocity, the acceleration at the end of the contraction, the stagnation pressure if the upstream static pressure is 1 bar, and the drop in static pressure across the contraction. Assume steady, one-dimensional, incompressible, frictionless flow.

(Answers: 77.17 m/s, 157.5 m/s, 4.25 × 10^4 m/s^2, 1.0197 bar, 6219 Pa)

7.3   Calculate the Bernoulli constant and the stagnation pressure at a location in a pipeline where the water velocity is 25 m/s, the static pressure 8 bar and the elevation 65 m.

(Answers: 17.5 bar, 11.125 bar)

7.4   For the flow in problem 7.3, calculate the pressure head, the dynamic head, the total head and the piezometric head. If the pipeline cross-sectional area is 1 m^2, what is the kinetic energy flowrate?

(Answers: 81.55 m, 31.86 m, 178.4 m, 146.6 m, 7.81 MW)

7.5   The exhaust gas from a turbojet engine has a density of 0.18 kg/m^3 and a soundspeed of 600 m/s. If the exhaust gas flowrate is 600 kg/s and the engine exhaust has a cross-sectional area of 4 m^2, calculate the velocity of the gas and the Mach number. Calculate the stagnation pressure of the air entering the engine if its static pressure is 0.5 bar, its density 0.7 kg/m^3 and the inlet area is 5 m^2. The mass flowrates of air and exhaust can be assumed to be the same.

(Answers: 833.3 m/s, 1.39, 0.603 bar)

CHAPTER 8

Engineering applications of Bernoulli's equation

Bernoulli's equation is so valuable in analysing a wide variety of fluid-flow problems that we now devote an entire chapter to illustrating how it is applied in practice, frequently together with the continuity equation. Instrumentation for flow measurement provides several application examples, including the Pitot-static tube for velocity measurement and the Venturi-tube and orifice-plate meters for the measurement of total fluid flowrate. We show how Bernoulli's equation can be used to give some insight into aerofoil lift and into the aerodynamic characteristics of a Formula One racing car. Another problem considered is that of liquid draining from a tank due to the influence of gravity. Yet another important application is shown to be the determination of the conditions for the onset of cavitation. Several examples are included which require Bernoulli's equation to be combined with aspects of hydrostatics for their solution.

8.1 Venturi-tube flowmeter

The Venturi tube, named after the Italian scientist Giovanni Battista Venturi (1746–1822), is one of a number of inline flowmeters, designed on the basis of Bernoulli's equation, which are commonly used to measure the total volumetric rate $\dot{Q}$ at which a low-viscosity gas or liquid flows through a pipe. A typical Venturi tube is illustrated in Figure 8.1. The essential features are a gradual conical contraction from the initial pipe diameter to a cylindrical throat followed by an even more gradual area increase usually back to the original pipe diameter. The internal geometry of the convergent first section of the Venturi tube, between the inlet and the throat, sometimes called a confuser, is intended to accelerate the fluid flowing through it and thereby reduce the fluid pressure. As we now show, the flowrate is derived from a measurement of the piezometric pressure drop across this upstream section. The absolute pressure is of importance only insofar as it influences the fluid density or the tendency for a liquid to cavitate (see section 8.10).

If we assume that the flow through the convergent section of the Venturi tube is one-dimensional, frictionless and the fluid has constant density ρ_F, we can apply Bernoulli's equation between the sections marked ① and ②

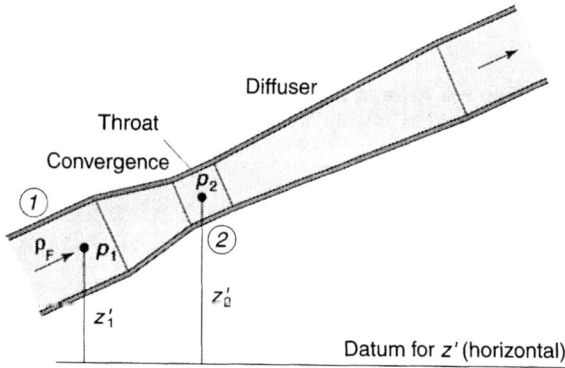

Figure 8.1 *Venturi-tube flowmeter*

in Figure 8.1

$$p_1 + \rho_F g z'_1 + \tfrac{1}{2} \rho_F V_1^2 = p_2 + \rho_F g z'_2 + \tfrac{1}{2} \rho_F V_2^2$$

which we can rearrange as

$$(p_1 + \rho_F g z'_1) - (p_2 + \rho_F g z'_2) = \tfrac{1}{2} \rho_F (V_2^2 - V_1^2).$$

If the Venturi tube is installed with its axis horizontal such that $z'_1 = z'_2$, the terms on the left-hand side reduce to the static pressure difference $\Delta p = p_1 - p_2$. More generally, however, a manometer or differential pressure transducer connected between sections ① and ② will measure the piezometric pressure difference ΔP

$$\Delta p + \rho_F g \, \Delta z' = \Delta P = \rho_F g \, \Delta H \qquad (8.1)$$

where $\Delta z'$ is the height difference between sections ① and ② and ΔH is the corresponding piezometric head difference.

In the event that either the static pressure difference Δp or the separate static pressures p_1 and p_2 are measured directly, the piezometric pressure difference ΔP is determined by adding $\rho_F g \, \Delta z'$ to Δp, i.e. the absolute height of the Venturi tube is of no significance, only the height difference $\Delta z'$.

We can now rewrite Bernoulli's equation in the convenient form

$$\Delta P = \tfrac{1}{2} \rho_F (V_2^2 - V_1^2). \qquad (8.2)$$

From the continuity equation

$$\dot{Q} = AV,$$

the velocities V_1 and V_2 can be written in terms of the volumetric flowrate $\dot{Q}$ and the cross-sectional areas at sections ① and ②, A_1 and A_2, as

$$V_1 = \frac{\dot{Q}}{A_1} \quad \text{and} \quad V_2 = \frac{\dot{Q}}{A_2}.$$

Substitution for V_1 and V_2 in equation (8.2) then gives

$$\Delta P = \tfrac{1}{2} \rho_F \, \dot{Q}^2 \left(\frac{1}{A_2^2} - \frac{1}{A_1^2} \right)$$

so that, after rearrangement, we have

$$\dot{Q} = A_2 \sqrt{\frac{2 \, \Delta P}{\rho_F [1 - (A_2/A_1)^2]}} \tag{8.3}$$

from which the volumetric flow rate $\dot{Q}$ can be determined. The corresponding expression for the mass flowrate $\dot{m} = \rho_F \dot{Q}$ is

$$\dot{m} = A_2 \sqrt{\frac{2 \rho_F \, \Delta P}{1 - (A_2/A_1)^2}} \tag{8.4}$$

Since $\Delta P / \rho_F = g \, \Delta H$, the expression for $\dot{Q}$ can be written in terms of the piezometric head difference ΔH as

$$\dot{Q} = A_2 \sqrt{\frac{2g \, \Delta H}{1 - (A_2/A_1)^2}} \tag{8.5}$$

8.2 Venturi-tube design and the coefficient of discharge

Equations (8.3), (8.4) and (8.5) are all based directly on Bernoulli's equation and the continuity equation for one-dimensional flow. The constant-density assumption is always valid for single-phase liquid flows and, as we showed in the previous chapter, remains very accurate even for gas velocities approaching the speed of sound. Any error associated with the constant-density assumption is further minimised by the fact that the fluid density ρ_F appears in the expressions for $\dot{Q}$ and $\dot{m}$ within the square root.

The influence on the accuracy of equations (8.3) and 8.4) of the one-dimensional flow assumption is far less easy to quantify than the constant-density assumption. The radial distribution of the fluid velocity for the flow upstream of the Venturi tube is likely to be far from uniform because any real flow is affected by viscosity. For this reason it is usual to calibrate Venturi tubes against a standard of very high accuracy to determine a performance factor called the *coefficient of discharge* C_D which is defined as the ratio of the actual flowrate $\dot{Q}_A$ to the theoretical flowrate based upon Bernoulli's equation $\dot{Q}_{TH}$, i.e.

$$C_D \equiv \frac{\dot{Q}_A}{\dot{Q}_{TH}} = \frac{\dot{Q}_A}{A_2} \sqrt{\frac{\rho_F[1 - (A_2/A_1)^2]}{2\,\Delta P}}$$

It should be recognised that the coefficient of discharge is a direct measure of the accuracy of the theory given in the preceding section. It is quite remarkable therefore that values of the coefficient of discharge C_D for low-viscosity liquid (e.g. water) flow through well-designed Venturi tubes can be as high as 0.995, suggesting that this very simple theory is almost perfect in this application. Calibration is normally carried out over a wide range of flow conditions and the results presented in the form of C_D versus the pipe-flow Reynolds number $\rho_F V_1 D_1/\mu_F$ where D_1 is the pipe diameter and μ_F the dynamic viscosity of the fluid.

The section of the Venturi tube downstream of the throat, which is known as a diffuser, has little if any influence on the characteristics of the Venturi tube as a flowmeter and is designed to minimise the total pressure loss involved in bringing the fluid velocity back to its upstream value. British Standard BS EN ISO 5167–1:1997: Part 1 specifies optimum values of 21° and of 7° to 8° for the convergence and divergence angles, respectively, for the two sections of the Venturi tube, of 0.3 to 0.75 for the overall throat-to-pipe diameter ratio, and also the installation requirements for a Venturi tube to give results within a known level of uncertainty. A key requirement is a run of straight undisturbed pipe (e.g. free of valves and bends) upstream of the Venturi tube at least 40 diameters in length (including a flow conditioner). As we discussed in section 7.6, total pressure loss represents a loss of mechanical energy and the product of flowrate times total pressure loss, $\dot{Q}\,\Delta p_T$, is the corresponding power required to maintain the flow against this pressure loss. It is evident from this consideration that the designer of a Venturi tube has to trade off the long-term operating costs associated with this power requirement against the capital cost of manufacturing the Venturi tube to achieve the lowest possible pressure loss through tight dimensional tolerances, smooth 'wetted'[*] surfaces and correct installation.

EXAMPLE 8.1

A Venturi-tube flowmeter installed in a horizontal pipe of diameter 80 mm has a throat diameter of 50 mm. The flowing fluid is compressed air with a density of 5 kg/m³ and a dynamic viscosity of 1.8×10^{-5} Pa.s. In a calibration test at a mass flowrate of 1.5 kg/s, the pressure upstream of the Venturi

---

[1] The word 'wetted' here means in contact with the flowing fluid, whether liquid or gas.

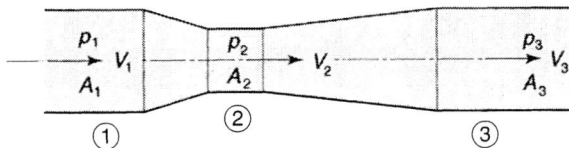

Figure E8.1

tube is found to be 4.20 bar, the throat pressure 3.69 bar, and the pressure at the Venturi-tube exit 4.19 bar. Calculate the coefficient of discharge, the pipe Reynolds number, the pressure-loss coefficient for the Venturi tube and the power dissipated by the fluid flowing through the Venturi tube.

SOLUTION
$D_1 = 0.08$ m; $D_2 = 0.05$ m; $D_3 = 0.08$ m; $\rho_F = 5$ kg/m³;
$\mu_F = 1.8 \times 10^{-5}$ Pa.s; $\dot{m}_A = 1.5$ kg/s; $p_1 = 4.20 \times 10^5$ Pa;
$p_2 = 3.69 \times 10^5$ Pa; $p_3 = 4.19 \times 10^5$ Pa

To find the coefficient of discharge C_D means that we need to find the theoretical flowrate $\dot{m}_{TH}$ given by equation (8.4) (in an examination, it is likely that this equation would have to be derived from Bernoulli's equation rather than simply remembered):

$$\dot{m}_{TH} = A_2 \sqrt{\frac{2\rho_F(p_1 - p_2)}{1 - (A_2/A_1)^2}}$$

$$= \frac{\pi \times 0.05^2}{4} \sqrt{\frac{2 \times 5 \times 5.1 \times 10^4}{1 - \left(\frac{0.05}{0.08}\right)^4}}$$

$$= 1.523 \text{ kg/s}$$

so that

$$C_D = \frac{1.5}{1.523} = 0.985$$

Since the pipe Reynolds number Re is found from

$$Re = \frac{\rho_F V_1 D_1}{\mu_F},$$

we need first to calculate the velocity V_1 from the continuity equation, i.e.

$$V_1 = \frac{\dot{m}_A}{\rho_F A_1} = \frac{1.5}{5 \times \pi \times (0.08^2/4)} = 59.7 \text{ m/s}$$

and so

$$Re = \frac{5 \times 59.7 \times 0.08}{1.8 \times 10^{-5}} = 1.33 \times 10^6.$$

The pressure-loss coefficient was defined in section 7.5 as

$$C_P = \frac{\Delta p}{\frac{1}{2}\rho_F V^2}$$

In this case we take $\Delta p = p_1 - p_3 = 10^3$ Pa, i.e. the pressure difference between two sections where the areas are the same, and $V = V_1$ so that

$$C_P = \frac{10^3}{0.5 \times 5 \times 59.7^2} = 0.112.$$

Finally, the power dissipated by the flow is given by

$$P = \frac{\dot{m}_A \Delta p}{\rho_F} = 300 \text{ W}.$$

Comment: While the value of C_D indicates a well-designed upstream section, the C_P value is rather high and this is reflected in the power dissipated which would be converted into thermal energy and result in a small rise in fluid temperature (about 0.05°C in this case).

8.3 Other Venturi-tube applications

The pressure reduction produced by flow through a Venturi tube leads to its use as a suction device in a number of practical applications including gas-fired water heater control systems, carburettors, and firehose foam injectors. A typical application of this type is illustrated in Example 8.2.

EXAMPLE 8.2
The arrangement shown in Figure E8.2 is used to inject liquid into a gas stream. If the stagnation pressure of the gas flow is p_0, show that the

minimum volumetric flowrate $\dot{Q}$ of gas through the convergent–divergent nozzle (Venturi tube) which will produce a liquid flow is given by

$$\dot{Q} = A\sqrt{[2(p_0 - B + \rho_L gH)/\rho_G]}$$

where ρ_G is the gas density, ρ_L the liquid density, A is the cross-sectional area of the Venturi throat, B is the barometric pressure (which acts on the liquid surface) and H is the vertical height of the injection tube tip above the liquid surface. Any effect of the injection tube on the gas flow can be neglected and the gas flow can be considered as loss free. The gas density can be neglected relative to the liquid density.

If the relationship between the mass flowrate of liquid $\dot{m}$ and the frictional pressure drop Δp_f between the injector tube tip and the inlet to the injector tube is

$$\dot{m} = C \, \Delta p_f$$

where $C = 1.33 \times 10^{-6}$ m.s, calculate the volumetric flowrate of gas required to produce a liquid mass flowrate of 8×10^{-3} kg/s. The gas density is 1.2 kg/m^3 and the stagnation pressure of the gas stream is 1.1 bar. The tip of the injector tube is 100 mm above the liquid surface and the liquid density is 800 kg/m^3. The throat area of the Venturi tube is 10^{-3} m^2 and the barometric pressure is 1.01 bar.

Convergent–divergent nozzle

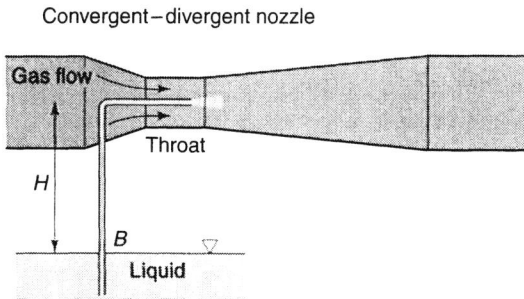

Figure E8.2

SOLUTION
According to Bernoulli's equation, as the gas flowrate $\dot{Q}$ through the Venturi tube is progressively increased, the static pressure p at the throat will drop such that

$$p_0 - p = \tfrac{1}{2} \rho_G V^2$$

where the velocity V of the gas in the throat section is obtained from the continuity equation

$$V = \frac{\dot{Q}}{A}.$$

These two equations can be combined to give

$$p_0 - p = \tfrac{1}{2} \rho_G \left(\frac{\dot{Q}}{A}\right)^2$$

Once the pressure p falls below the barometric pressure B, the injection tube will act much like a piezometer tube and the liquid will rise to a height h above the level of the liquid in the reservoir given by the hydrostatic equation

$$p + \rho_L g h = B.$$

Once the liquid has risen to the top of the injection tube, such that $h = H$, it will begin to flow into the gas stream. The corresponding value for the pressure p is given by

$$p + \rho_L g H = B \quad \text{or} \quad p = B - \rho_L g H$$

and, by substituting for p in the equation connecting $p_0 - p$ with $\dot{Q}$, we have

$$p_0 - B + \rho_L g H = \tfrac{1}{2} \rho_G \left(\frac{\dot{Q}}{A}\right)^2.$$

After rearrangement, this equation gives

$$\dot{Q} = A \sqrt{[2(p_0 - B + \rho_L g H)/\rho_G]}.$$

which corresponds to the value of $\dot{Q}$ which must be exceeded to produce a flow of liquid into the gas stream.

Higher volumetric flowrates than the minimum corresponding to the equation for $\dot{Q}$ will result in a liquid mass flowrate $\dot{m}$ according to

$$\dot{m} = C \, \Delta p_f$$

where Δp_f is the pressure drop over the length of the injection tube associated with the viscosity of the liquid. The overall pressure drop for the injection tube is the sum of Δp_f and the hydrostatic pressure difference $\rho_L g H$, i.e.

$$B - p = \Delta p_f + \rho_L g H.$$

For the numerical part of the problem we have

$C = 1.33 \times 10^{-6}$ m.s; $\dot{m} = 8 \times 10^{-3}$ kg/s; $\rho_G = 1.2$ kg/m³;
$p_0 = 1.1 \times 10^5$ Pa; $H = 0.1$ m; $\rho_L = 800$ kg/m³;
$A = 10^{-3}$ m²; $B = 1.01 \times 10^5$ Pa

To produce a liquid flowrate of 8×10^{-3} kg/s requires

$$\Delta p_f = \frac{\dot{m}}{C} = \frac{8 \times 10^{-3}}{1.33 \times 10^{-6}} = 6015 \text{ Pa}$$

so that

$$B - p = 6015 + 800 \times 9.81 \times 0.1 = 6800 \text{ Pa}.$$

Since we have $B = 1.01 \times 10^5$ Pa, we find $p = 9.42 \times 10^4$ Pa.
The relationship between p_0, p and $\dot{Q}$ is still valid, i.e.

$$p_0 = p + \tfrac{1}{2} \rho_G \left(\frac{\dot{Q}}{A}\right)^2.$$

Since the stagnation pressure p_0 is given as 1 bar, we have

$$\left(\frac{\dot{Q}}{A}\right)^2 = \frac{2(p_0 - p)}{\rho_G} = \frac{2 \times 5800}{1.2} = 9666 \text{ m}^2/\text{s}^2$$

from which the flow rate $\dot{Q}$ is

$$\dot{Q} = 10^{-3} \times \sqrt{9666} = 0.0983 \text{ m}^3/\text{s}.$$

The corresponding gas velocity in the throat is 98.3 m/s which is well below the level at which compressibility effects become significant.

8.4 Orifice-plate flowmeter

A relatively simple and therefore cheap alternative to the Venturi-tube flowmeter is the orifice-plate flowmeter. In essence an orifice plate is a thin disk with a hole (the orifice) in it which has an open area smaller than that of the pipe cross-section. In principle the orifice can be of any shape and located anywhere in the disk but is usually a circular hole, concentric with the pipe bore, in the diameter range $0.75D_1 > D_2 > 0.20D_1$ (minimum value 12.5 mm), with sharp bevelled edges. In a typical installation, as shown in Figure 8.2, the orifice plate is held in place between two flanges and the pressure drop which results from the acceleration of the fluid passing through the orifice plate is measured between pressure tappings located at distances D_1 upstream and $D_1/2$ downstream of the

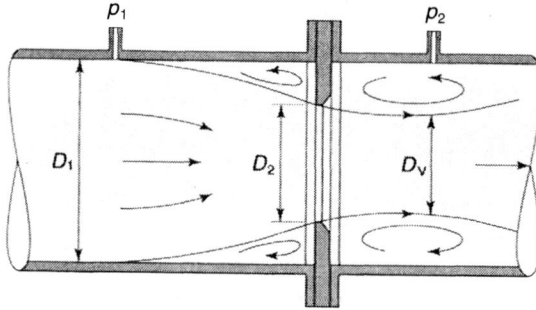

Figure 8.2 *Orifice-plate flowmeter*

orifice plate. A full specification for the design and installation of orifice-plate flowmeters is given in BS EN ISO 5167–1: 1997.

The basic analysis of flow through an orifice is identical to that for a Venturi tube with the final result for the theoretical volumetric flowrate in terms of the pressure difference $p_1 - p_2$, the fluid density ρ_F and the areas A_1 and A_2 of the pipe cross-section and the orifice, respectively, i.e.

$$\dot{Q}_{TH} = A_2 \sqrt{\frac{2(p_1 - p_2)}{\rho_F[1 - (A_2/A_1)^2]}} \tag{8.3}$$

It turns out that the value of the coefficient of discharge $C_D = \dot{Q}_A/\dot{Q}_{TH}$, even for the most carefully designed orifice-plate flowmeter, is about 0.6. Such a low value suggests that, in contrast with the Venturi-tube situation, Bernoulli's equation is a poor basis on which to analyse flow through an orifice. Reference to Figure 8.2 reveals that the fault is not so much with Bernoulli's equation but with the way it has been applied. In effect, the fluid passing through an orifice creates its own Venturi tube with a contraction section starting at about location 1 and a throat at location 2. This fluid throat is called the *vena contracta* (which means the contracted vein) and if its cross-sectional area A_V were substituted in equation (8.3) instead of A_2, the result would be a coefficient of discharge practically equal to one. Unfortunately the diameter of the *vena contracta* is not a fixed quantity which we can easily measure and in practice orifice plates are widely used with coefficients of discharge based either upon a standard or determined from calibration tests. It is important to understand that a low value for C_D does not mean that orifice-plate flowmeters are inherently inaccurate: the accuracy is determined by the value of C_D.

Immediately upstream and downstream of the orifice plate the central stream of high-velocity fluid is surrounded by recirculating eddies of fluid within which the fluid velocity is relatively low (sometimes called a deadwater zone) and there is minimal pressure variation. Location 2, a distance D_1 downstream of the orifice plate, corresponds with the position of minimum pressure (the *vena contracta*) so that the pressure difference measured is as high as possible to improve accuracy. The loss of total pressure, i.e. the irrecoverable pressure loss, is much higher for an orifice plate than for a Venturi tube, in part due to the contraction region but primarily as a result of the rather violent way in which the flow recovers downstream of the orifice without the aid of a diffuser. As with the Venturi tube, there is an economic trade off between the low cost of an orifice plate and the operating costs associated with the irrecoverable pressure loss.

EXAMPLE 8.3

An orifice-plate flowmeter with an orifice diameter of 350 mm and a coefficient of discharge of 0.6 is used to monitor the flowrate of water in a pipe of diameter 500 mm. Calculate the volumetric flowrate if the pressure difference across the orifice plate is 1.26 bar and calculate the diameter of the *vena contracta*.

SOLUTION

$D_2 = 0.35$ m; $D_1 = 0.5$ m; $C_D = 0.6$; $p_1 - p_2 = 1.26 \times 10^5$ Pa; $\rho_F = 1000$ kg/m^3

We have

$$A_1 = \tfrac{1}{4} \pi D_1^2 = 0.196 \text{ m}^2$$

$$A_2 = \tfrac{1}{4} \pi D_2^2 = 0.0962 \text{ m}^2$$

$$\frac{A_2}{A_1} = \left(\frac{0.35}{0.5}\right)^2 = 0.49.$$

From equation (8.3)

$$\dot{Q}_{TH} = 0.0962 \sqrt{\frac{2 \times 1.26 \times 10^5}{10^3 \times (1 - 0.49^2)}} = 1.75 \text{ m}^3/\text{s}$$

and so

$$\dot{Q}_A = C_D \dot{Q}_{TH} = 1.05 \text{ m}^3/\text{s}.$$

The diameter of the *vena contracta* D_V corresponds to the value of A_2 in equation (8.3) which results in the actual volumetric flowrate, i.e.

$$\dot{Q}_A = A_V \sqrt{\frac{2(p_1 - p_2)}{\rho_F[1 - (A_V/A_1)^2]}}$$

which can be rearranged to give

$$\frac{1}{A_V^2} = \frac{2(p_1 - p_2)}{\rho_F \dot{Q}_A^2} + \frac{1}{A_1^2}.$$

Substitution of the values for $p_1 - p_2$, ρ_F, $\dot{Q}_A$ and A_1 leads to

$$\frac{1}{A_V^2} = \frac{2 \times 1.26 \times 10^5}{10^3 \times 1.05^2} + \frac{1}{0.196^2}$$

from which $A_V = 0.0627$ m^2 and $D_V = 0.283$ m or 28.3 mm, i.e. the diameter of the *vena contracta* is about 19% smaller than that of the orifice.

8.5 Aerodynamic lift

The cross-sectional shape of an aircraft wing (also called an aerofoil), such as that shown in Figures 1.6 and 6.1, is designed so that the airflow over the upper surface leads to pressures which are on average lower than the pressures which result from the airflow over the lower surface. The product of the difference between the average pressures and the planform area of the wing A corresponds to the lift force L produced by the wing. It is usual to present lift data in the form of a lift coefficient C_L defined by

$$C_L = \frac{L}{\frac{1}{2}\rho V_0^2 A}$$

where ρ is the density and V_0 the relative airspeed of the airflow upstream of the wing. For any given wing, the lift coefficient depends mainly on the angle of attack α, usually increasing with α to a maximum and then falling sharply as the wing stalls due to separation of the boundary layer (discussed briefly in section 6.4) on the suction (i.e. upper) surface.

The calculation of lift from first principles, given only the aerofoil shape and fluid properties, is well beyond the scope of this text. However, as we show in the following example, if the velocity variation is known we can use Bernoulli's equation to calculate the pressure variation over the surface of the wing and hence the lift coefficient.

EXAMPLE 8.4

The variation of velocity V for the flow of air over the upper surface of a stationary aerofoil of chord length C is given by

$$\frac{V}{V_0} = \frac{17}{2}\frac{x}{C} - \frac{15}{2}\left(\frac{x}{C}\right)^2$$

where x is the distance from the leading edge of the aerofoil and V_0 is the velocity of the flow far from the aerofoil. The air velocity over the lower surface may be taken as constant and equal to V_0. Derive an expression for the variation with x of the pressure difference between the lower and upper surfaces of the aerofoil. Show that the lift force resulting from this pressure difference is given by

$$L = \tfrac{59}{48}\, \rho V_0^2 S C$$

where S is the span of the aerofoil and ρ is the air density. Assume the flow is incompressible, frictionless and unaffected by gravity.

If the air velocity V_0 is 100 m/s and its density is 1.2 kg/m³, calculate the minimum pressure difference between the upper and lower surfaces of the aerofoil at any x location. Calculate the lift force on an aerofoil of span 20 m and chord 1 m. If the stagnation pressure of the flow is 1.05 bar, calculate the force exerted by the air on the lower surface of the aerofoil.

SOLUTION
Bernoulli's equation for the upper surface may be written as

$$p_0 = p_U + \tfrac{1}{2}\, \rho V^2$$

and for the lower surface as

$$p_0 = p_L + \tfrac{1}{2}\, \rho V_0^2$$

where the subscripts U and L refer to the upper and lower surfaces respectively.

The pressure difference $p_L - p_U$ is thus given by

$$p_L - p_U = \tfrac{1}{2}\, \rho (V^2 - V_0^2)$$

$$= \tfrac{1}{2}\, \rho V_0^2 \left[\left(\frac{V}{V_0} \right)^2 - 1 \right].$$

If we define $X \equiv x/C$, we can now substitute for V/V_0 terms of X to find

$$p_L - p_U = \tfrac{1}{2}\, \rho V_0^2 \{ [\tfrac{17}{2} X - \tfrac{15}{2} X^2]^2 - 1 \}.$$

The lift force is given by integrating this pressure difference over the plan-form area, i.e.

$$L = \int_0^C (p_L - p_U) S \; dx$$

$$= \tfrac{1}{2}\, \rho V_0^2 S C \int_0^1 \{ [\tfrac{17}{2} X - \tfrac{15}{2} X^2]^2 - 1 \} \; dX$$

$$= \tfrac{1}{2}\, \rho V_0^2 S C [(\tfrac{17}{2})^2 \times \tfrac{1}{3} - 2 \times \tfrac{17}{2} \times \tfrac{15}{2} \times \tfrac{1}{4} + (\tfrac{15}{2})^2 \times \tfrac{1}{5} - 1]$$

$$= \tfrac{59}{48}\, \rho V_0^2 S C.$$

To calculate the value of the minimum pressure difference, we must first find the x-location of the minimum. Since we have

$$p_L - p_U = \tfrac{1}{2}\rho V_0^2\{[\tfrac{17}{2}X - \tfrac{15}{2}X^2]^2 - 1\}$$

it should be apparent that $p_L - p_U$ will be a minimum when the term within the square brackets is a minimum (i.e. when the velocity V is a maximum). For $\tfrac{17}{2}X - \tfrac{15}{2}X^2$ to be a minimum requires that its derivative equals zero, i.e.

$$\tfrac{17}{2} - 2 \times \tfrac{15}{2}X = 0$$

from which

$$X = \frac{x}{C} = \tfrac{17}{30}.$$

For the numerical part of the problem we have

$V_0 = 100$ m/s; $\rho = 1.2$ kg/m^3; $S = 20$ m; $C = 1$ m;
$p_0 = 1.05 \times 10^5$ Pa

At $x/C = \tfrac{17}{30}$ we have

$$p_L - p_U = 0.5 \times 1.2 \times 100^2\{[\tfrac{17}{2} \times \tfrac{17}{30} - \tfrac{15}{2}(\tfrac{17}{30})^2]^2 - 1\}$$
$$= 2.9 \times 10^4 \text{ Pa} \quad \text{or} \quad 0.29 \text{ bar.}$$

The lift force is easily obtained as

$$L = \tfrac{59}{48} \times 1.2 \times 100^2 \times 20 \times 1$$
$$= 2.95 \times 10^5 \text{ N} \quad \text{or} \quad 0.295 \text{ MN.}$$

Since the stagnation pressure $p_0 = 1.05$ bar, the static pressure p_L on the lower surface can be obtained from Bernoulli's equation as

$$p_L = p_0 - \tfrac{1}{2}\rho V_0^2$$
$$= 1.05 \times 10^5 - \tfrac{1}{2} \times 1.2 \times 100^2$$
$$= 9.9 \times 10^4 \text{ Pa} \quad \text{or} \quad 0.99 \text{ bar}$$

and the magnitude of the force exerted on the lower surface is

$$p_L SC = 1.98 \times 10^6 \text{ N} \quad \text{or} \quad 1.98 \text{ MN.}$$

8.6 **Formula One racing car**

The modern Formula One, or Grand Prix, racing car is a complex package of mechanical and electronic components constructed to a formula (hence the name) defined by the Federation Internationale de l'Automobile (FIA) which prescribes a wide range of design parameters. Aerodynamic performance has long been a critical aspect of the design of a Formula One car and extensive wind-tunnel testing of large-scale models (50% of full size is not uncommon) is an essential element of racing-car development. Indeed, most leading teams now operate their own sophisticated wind tunnels which incorporate such special features as a rolling road to properly simulate the aerodynamic interaction between a car and the road over which it travels and cryogenic systems to achieve Reynolds numbers close to those which correspond to typical racing speeds (i.e. up to about 365 kph).

Figure 8.3 shows a highly idealised picture of a Formula One racing car. Front and rear aerofoils, designed to produce download (i.e. negative lift) on the front and rear wheels and thereby improve traction, are the most obvious aspects of the design motivated solely by aerodynamic considerations. Considerable download is also generated by the underside of the car which has a Venturi-like shape to reduce the pressure of air flowing under the car. Just as for the wings and other lifting surfaces of aircraft, devices which generate aerodynamic download inevitably result in drag (known as induced drag) to add to the drag associated with the exposed tyres, bodywork, radiators, oil coolers, engine inlet and the driver. The complexity of the aerodynamic problem is made even worse by the interaction between these components and, in actual racing, other cars – particularly when one car is travelling in the wake of another car a short distance in front. It is the intense trailing vortices, often visible in humid or damp conditions swirling away from the endplates on either side of the rear wing, which are responsible for the loss of download experienced by the following car. The same phenomenon affects one aircraft following another and can lead to catastrophic results for small aircraft following much larger aircraft such as jumbo jets.

On the basis of the material covered in Chapters 3, 6 and 7, we can make crude estimates of some aspects of the aerodynamic performance of a Formula One

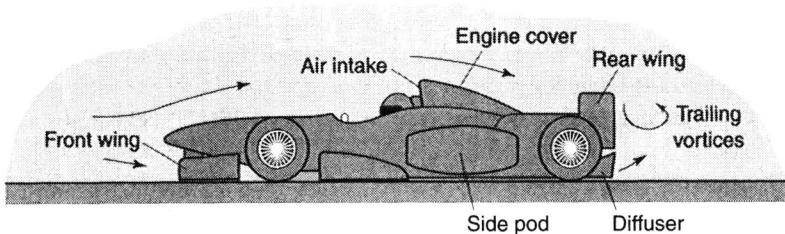

Figure 8.3 *Idealised Formula One racing car*

car. A more complete analysis would be immensely complicated and require knowledge of closely guarded design data.

We assume the following values, which have been estimated from published information:

Maximum speed 320 kph	$V = 88.9$ m/s
Tractive power 800 hp	$P = 597$ kW
Projected frontal area	$A_{FR} = 1.5$ m^2
Air density	$\rho = 1.2$ kg/m^3
Area reduction for flow beneath car	1.15 : 1
Mass of car and driver	$m = 700$ kg
Projected plan area	$A_{PL} = 7$ m^2

At maximum speed V, we assume that the tractive power P is used to overcome the aerodynamic drag D, so that

$$P = DV$$

and we can therefore calculate the aerodynamic drag to be

$$D = \frac{P}{V} = \frac{5.97 \times 10^5}{88.9} = 6716 \text{ N}.$$

In reality, the tractive power would also have to overcome rolling resistance but this force is certainly negligible compared with D.

We can now calculate the overall drag coefficient from

$$C_D = \frac{D}{\frac{1}{2} \rho V^2 A_{FR}}$$

$$= \frac{6716}{0.5 \times 1.2 \times 88.9^2 \times 1.5} = 0.944.$$

This value for C_D is about three times higher than the value one would expect for a well-designed passenger car for which low drag is desirable in order to reduce fuel consumption and noise. Neither of these considerations is of paramount importance to the designer of a Formula One car and the high value of C_D is a direct consequence of the induced drag associated with the very high levels of download.

If we assume the airflow under the car is loss free, we can apply Bernoulli's equation to estimate the reduction in pressure below the ambient level B, i.e.

$$B + \tfrac{1}{2} \rho V^2 = p_2 + \tfrac{1}{2} \rho V_2^2$$

where p_2 is taken as the pressure beneath the car and V_2 the corresponding airspeed. There are many things that can be criticised about this simple approximation. For example, the area beneath the car is far from constant, especially towards the rear where a diffuser brings the flow back to ambient pressure, and in fact the flow is likely to be three dimensional. However, Bernoulli's equation incorporates much of the essential physics of many flows and is unlikely to produce answers which are orders of magnitude different from reality.

For an area ratio of $1.15:1$, the continuity equation tells us that $V_2 = 1.15V = 102.2$ m/s and so

$$B - p_2 = 0.5 \times 1.2 \times (102.2^2 - 88.9^2) = 1529 \text{ Pa.}$$

If we now assume that this pressure difference is distributed uniformly over the projected plan area A_{PL}, the corresponding download is 10.7 kN which we can compare with the weight of the car $mg = 6.4$ kN. The outcome of this calculation is critically dependent upon the assumption about the area change, by which we effectively specify the volumetric flowrate of air under the car. The overall download is certainly well in excess of the weight of the car and our calculation suggests that the contribution due to the underflow may be substantial.

We conclude this section by estimating the retardation due to aerodynamic drag. If all tractive force (whether due to the engine or brakes) is lost, according to Newton's second law of motion we have

$$ma = -D.$$

Our estimate for the drag force D at maximum speed was 6716 N which leads to

$$-a = \frac{D}{m} = \frac{6716}{700} = 9.59 \text{ m/s}^2 \qquad \text{or} \qquad 0.98g$$

i.e. a deceleration close to $1g$ solely due to aerodynamic drag, a value well in excess of the braking capability of the majority of passenger cars.

8.7 Pitot tube

The simple L-shaped tube shown in Figure 8.4(a) is called a Pitot tube after the French hydraulic engineer Henri de Pitot (1695–1771) who devised this instrument in 1732 to measure the speed of a boat or of water flowing along an open channel or river. When immersed in a liquid flow to a depth Z as shown, the liquid enters the tube and rises to a level H above the free surface. If the flow is steady, once this equilibrium situation is reached the liquid velocity within the Pitot tube is zero and the point P at its tip becomes a stagnation point at which the pressure is equal to the stagnation pressure p_0. If V is the liquid velocity and p the static pressure at point O on the horizontal streamline which terminates at P, the point O being far enough upstream of the Pitot tube for conditions there not to be affected by its presence, then from Bernoulli's equation we have

$$p_0 = p + \tfrac{1}{2}\rho V^2 \quad \text{or} \quad p_0 - p = \tfrac{1}{2}\rho V^2.$$

From the hydrostatic equation (4.6) for a constant-density fluid, we have

$$p_0 = B + \rho g(H + Z) \quad \text{and} \quad p = B + \rho g Z$$

where B is the pressure acting on the liquid surface. If we eliminate B between these two equations, we have

$$p_0 - p = \rho g H$$

so that

$$V = \sqrt{(2gH)}.$$

The more usual arrangement for the measurement of the velocity of a flowing gas or liquid is to use a Pitot tube in combination with a probe to measure the static pressure p in the vicinity of the point P. As shown in Figure 8.4(b) for liquid flow in a pipe, the Pitot tube can be used together with a piezometer tube (see section 4.6). We now have

$$p = B + \rho g(h + R) \quad \text{and} \quad p_0 = B + \rho g(H + R),$$

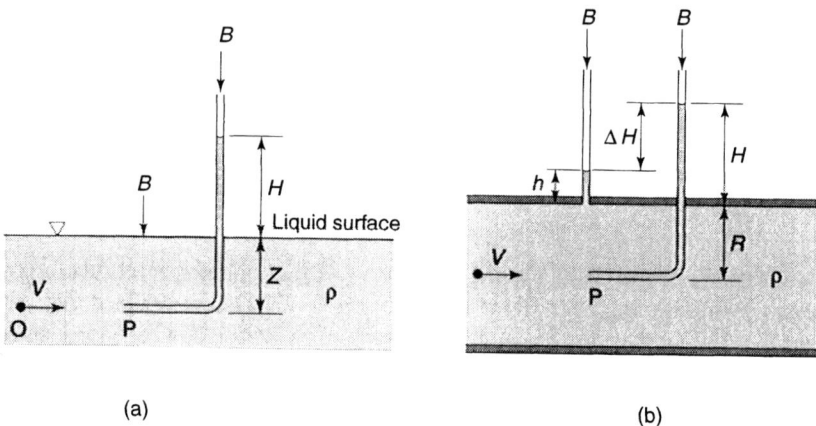

(a)

(b)

Figure 8.4 Pitot tube: (a) free surface flow; (b) pipe flow

where R is the radius of the pipe. From Bernoulli's equation we again have

$$p_0 - p = \tfrac{1}{2}\rho V^2$$

so that now

$$\tfrac{1}{2}\rho V^2 = \rho g(H - h) = \rho g\,\Delta H$$

and the velocity is given by

$$V = \sqrt{\frac{2\,\Delta p}{\rho}} = \sqrt{(2g\,\Delta H)}$$

where $\Delta p = p_0 - p$ and $\Delta H = H - h$. The first of these two equations for V applies to both liquids and gases.

There is an important difference between the application of Bernoulli's equation for the analysis of flow through flowmeters, such as the Venturi tube and the orifice plate, and to the analysis of a velocity probe immersed in a flowing fluid. For the flowmeters the analysis deals with all the fluid passing through a duct and so makes use not only of Bernoulli's equation but also the continuity equation and therefore rests upon the assumption that the flow is one dimensional. Since there is no flow through a Pitot tube, the analysis does not depend upon the one-dimensional approximation and requires only that we know the difference between the stagnation and static pressures at the measurement location.

EXAMPLE 8.4

The output from a differential pressure transducer, with one side connected to a Pitot tube immersed in a gas flow and the other side connected to a static-pressure tapping in the near vicinity of the Pitot tube, is 3.9 kPa. If the gas has a density of 0.8 kg/m^3 and a soundspeed of 330 m/s, calculate the gas velocity and the corresponding Mach number. If the transducer were to be replaced with a U-tube manometer, would kerosene (density 800 kg/m^3) or mercury be the more suitable manometer liquid?

SOLUTION

$p_0 - p = 3900 \text{ Pa};\quad \rho = 0.8 \text{ kg/m}^3;\quad c = 330 \text{ m/s};$
$\rho_M = 800 \text{ kg/m}^3 \text{ or } 13.6 \times 10^3 \text{ kg/m}^3$

We start with the result obtained above (this would have to be derived from Bernoulli's equation in an examination)

$$V = \sqrt{\frac{2(p_0 - p)}{\rho}} = \sqrt{\frac{2 \times 3900}{0.8}} = 98.7 \text{ m/s}.$$

The corresponding Mach number is

$$M = \frac{V}{c} = \frac{98.7}{330} = 0.299$$

i.e. the assumption of incompressible flow is just valid.

As we found in section 4.7, for a U-tube manometer we have

$$\Delta p = (\rho_M - \rho)g \, \Delta H.$$

For mercury we thus find

$$\Delta H = 0.0292 \text{ m} \quad \text{or} \quad 29.2 \text{ mm}$$

and for kerosene

$$\Delta H = 0.497 \text{ m} \quad \text{or} \quad 497 \text{ mm.}$$

On the basis of the height difference, either liquid would be suitable for the measurement. Kerosene would probably be the preferred choice if lower flowspeeds, which would generate smaller pressure and manometer level differences, were also to be covered provided this inflammable fluid was not ruled out on safety grounds.

8.8 Pitot-static tube

The arrangement illustrated in Figure 8.5, known as a Pitot-static tube, consists of an inner tube to sense the stagnation pressure p_0 of a flow and a concentric outer tube, of outer diameter D, closed at its upstream end but perforated by a series of small holes to sense the static pressure p of the flow. The static-pressure holes should be located sufficiently far downstream of the probe tip for any disturbance to the flow created by the probe to have died out: a distance of about $6D$ is found to be sufficient and about double that to any bend in the two tubes. To ensure an accurate measurement of static pressure, the Pitot-static tube must be aligned with the flow to within about 5°.

If the density of the fluid is ρ_F and the fluid velocity just upstream of the probe tip is V, then from Bernoulli's equation we have

$$p_0 = p + \frac{1}{2}\rho_F V^2 \quad \text{so that} \quad V = \sqrt{\frac{2(p_0 - p)}{\rho_F}}.$$

In a typical application, as shown in Figure 8.5, the pressure difference $p_0 - p$ is

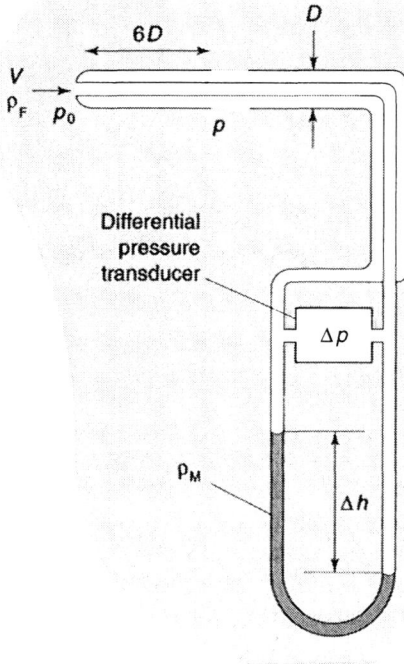

Figure 8.5 *Pitot-static tube*

measured using either a differential pressure transducer or a U-tube manometer. In the latter case, if the density of the manometer liquid is ρ_M and the vertical height difference due to the pressure difference is Δh, then we have

$$p_0 - p = (\rho_M - \rho_F)g\,\Delta h = \tfrac{1}{2}\rho_F V^2$$

from which

$$V = \sqrt{\frac{2(\rho_M - \rho_F)g\,\Delta h}{\rho_F}}.$$

8.9 Liquid draining from a tank

Figure 8.6 shows a container of cross-sectional area A_S open to atmospheric pressure B and containing a liquid of density ρ_F. Under the influence of gravity,

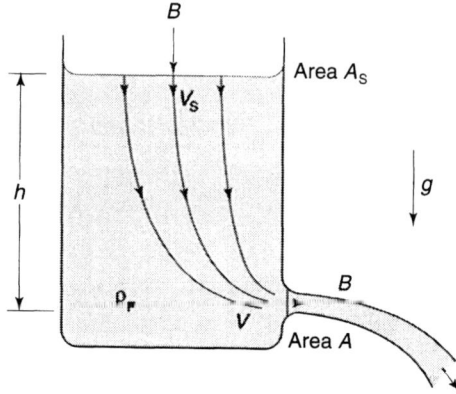

Figure 8.6 *Liquid draining from a tank*

the liquid drains out of the tank through an orifice in the tank wall of cross-sectional area A. In practice, the orifice could be simply a hole, a nozzle such as a Venturi tube, or a control valve.

To determine the volumetric flowrate $\dot{Q}$ with which the liquid flows out of the tank, we start by writing Bernoulli's equation for a streamline connecting the liquid surface and the jet emerging from the orifice:

$$p_T = B + \rho_F g h + \tfrac{1}{2}\rho_F V_S^2 = B + \tfrac{1}{2}\rho_F V^2$$

where p_T is the total pressure and h is the vertical height of the liquid surface above the centre of the orifice, V is the velocity of the liquid passing through the orifice and V_S is the surface velocity. An unusual feature of this problem is that the static pressure of the liquid is equal to the surrounding (ambient) pressure both at the liquid surface and also at the location of the orifice.

The continuity equation in this case is as follows

$$\dot{Q} = V_S A_S = VA.$$

In a typical situation, the surface area A_S is far greater than the orifice area A so that $V \gg V_S$ and we may neglect the term $\tfrac{1}{2}\rho_F V_S^2$ in comparison with $\tfrac{1}{2}\rho_F V^2$ in Bernoulli's equation, which then simplifies to

$$V^2 = 2gh \quad \text{or} \quad V = \sqrt{(2gh)}$$

which is called Torricelli's formula.

The theoretical volumetric flowrate $\dot{Q}_{TH}$ can now be written as

$$\dot{Q}_{TH} = VA = A\sqrt{(2gh)}$$

and the actual flowrate $\dot{Q}_A$ as

$$\dot{Q}_A = C_D A \sqrt{(2gh)}$$

where C_D is a coefficient of discharge for flow through the orifice.

It is now straightforward to calculate the time t required for the liquid level to fall to h from an initial value h_0 (i.e. at $t = 0$). We make use of the kinematic relation

$$V_S = -\frac{dh}{dt}$$

i.e. the downward velocity of the liquid surface must equal the rate of change of the liquid level h. We can combine this result with the continuity equation and the expression for the volumetric flowrate $\dot{Q}_A$ as follows:

$$\dot{Q}_A = A_S V_S = -A_S \frac{dh}{dt} = C_D A \sqrt{(2gh)}$$

which yields a first-order differential equation for h as a function of time, i.e.

$$h^{-\frac{1}{2}} \frac{dh}{dt} = -C_D \frac{A}{A_S} \sqrt{(2g)}.$$

This equation can be integrated to give the desired relationship between liquid level h and the time t

$$2(\sqrt{h} - \sqrt{h_0}) = -C_D \frac{A}{A_S} t \sqrt{(2g)}$$

wherein the constant of integration has been determined from the initial condition $h = h_0$ at $t = 0$.

It should be evident that as the liquid flows out of the tank, both the level h and the liquid flowrate $\dot{Q}$ will decrease. What this means is that we have been dealing with an unsteady flow problem yet we have analysed it on the basis of Bernoulli's equation and the continuity equation, both of which were derived assuming steady flow. To see whether the assumption of steady flow is justified, we need to compare an estimate of the unsteady acceleration term dV/dt with the acceleration along the streamline $V(dV/ds)$.

From $V^2 = 2gh$ we have

$$\frac{dV}{dt} = \frac{g}{V} \times \frac{dh}{dt} = \frac{-gV_S}{V} = \frac{-gA}{A_S}.$$

It is less obvious how we might estimate the term $V(dV/ds)$. One possibility is to recognise that between the surface and the orifice the fluid velocity increases from V_S to V over a distance comparable with h. In the absence of a much more detailed analysis, we cannot be more precise than this. We note that at any time this distance is greater than h because the streamline follows a curved path but that most of the velocity change occurs in the vicinity of the orifice. On this basis the acceleration can be estimated as $V(V/h) = V^2/h = 2g$. Although this is a very crude analysis, it indicates that, provided $A_S/A \gg 1$, the steady flow assumption is well justified.

As the following example illustrates, the foregoing analysis can be extended without great difficulty to the situation where the tank is closed and a pressure p_R is imposed on the liquid surface.

EXAMPLE 8.5

Gas at a pressure p_R is used to force a liquid of density ρ out of a container, as shown in Figure E8.5. The liquid leaves the container through a Venturi tube of exit area A. Show that the mass flow rate $\dot{m}$ of the liquid is given by

$$\dot{m} = A\sqrt{[2\rho(p_R - B + \rho gh)]}$$

where h is the vertical height of the liquid surface above the Venturi tube. The flow may be assumed to be steady, frictionless and one dimensional and the liquid pressure at exit from the Venturi tube equal to that of the surrounding atmosphere B. The downward velocity of the liquid surface may be regarded as negligible compared with that of the liquid jet.

If the liquid has a density of $800 \ kg/m^3$ and its surface is a height above the Venturi tube of 3 m, the applied pressure is 2 bar, the nozzle exit area is $0.01 \ m^2$, and the atmospheric pressure is 1.01 bar, calculate the liquid

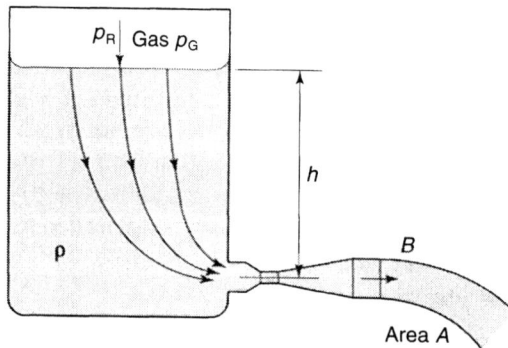

Figure E8.5

mass flowrate. Calculate also the jet velocity and the velocity of the liquid surface if the cross-sectional area of the tank is 1 m^2.

SOLUTION
Since the flow can be treated as steady, frictionless and one dimensional, we can apply Bernoulli's equation between the liquid surface and the Venturi tube exit, as follows:

$$p_T = p_R + \rho gh + \tfrac{1}{2}\,\rho V_S^2 = B + \tfrac{1}{2}\,\rho V^2$$

where p_T is the total pressure, which is constant along a streamline, and V_S is the surface velocity. We can rearrange the equation as follows:

$$V^2 - V_S^2 = \frac{2(p_R - B + \rho gh)}{\rho}.$$

Since V_S can be neglected with respect to V, we have

$$V = \sqrt{\frac{2(p_R - B + \rho gh)}{\rho}}.$$

We now introduce the continuity equation

$$\dot{m} = \rho AV$$

and substitute for V to obtain

$$\dot{m} = A\sqrt{[2\rho(p_R - B + \rho gh)]}.$$

For the numerical part of the problem we have

$p_R = 2 \times 10^5 \text{ Pa};\quad B = 1.01 \times 10^5 \text{ Pa};\quad \rho = 800 \text{ kg/m}^3;$
$h = 3 \text{ m};\quad A = 0.01 \text{ m}^2;\quad A_S = 1 \text{ m}^2$

It is a matter of straightforward substitution to find

$$\dot{m} = 140 \text{ kg/s}.$$

From the continuity equation

$$V = \frac{\dot{m}}{\rho A} \quad \text{and} \quad V_S = \frac{\dot{m}}{\rho A_S}$$

so that

$$V = \frac{140}{800 \times 0.01} = 17.5 \text{ m/s} \quad \text{and} \quad V_s = \frac{140}{800 \times 1} = 0.175 \text{ m/s}.$$

We shall return to this problem at the end of the next section.

8.10 Cavitation in liquid flows

It should be clear from Bernoulli's equation that in a flowing liquid it is possible to develop very low pressure in regions of high liquid velocity. If this pressure falls below the saturated vapour pressure, tiny vapour bubbles begin to form, a process known as cavitation. For a given temperature, the saturated vapour pressure is the pressure at which a liquid boils and is in equilibrium with its own vapour, i.e. it is the pressure which exists in pure vapour in contact with the liquid at a given temperature. The variation of the saturated vapour pressure with temperature for water, shown in Figure 2.7 and tabulated in Table A.2 of the appendix, is based upon the saturation table for water and steam (see Mayhew and Rogers 1980). As we should expect, the saturated vapour pressure at 100°C is 1.01 bar, i.e. at normal atmospheric pressure water boils at 100°C. If the pressure is reduced to 0.1 bar, water boils at 45.8°C whereas for a pressure of 20 bar the boiling point is raised to 212.4°C. An application which takes advantage of the influence of pressure on the boiling point is the domestic pressure cooker.

Vapour bubbles formed due to the pressure reduction in a flowing liquid initially grow, are swept downstream and then collapse implosively upon reaching a zone of sufficiently high pressure. Cavitation in pumps and hydraulic turbines is undesirable, first because it leads to a decrease in efficiency and secondly because repeated impacts on blading and other components, due to the collapse of the vapour bubbles, can be so intense as to cause serious wear and other mechanical damage. Much the same is true for ships' propellers where cavitation can occur at the propeller tips. Further cavitation examples are provided by nozzles, valves and pipes where there are no moving parts but the liquid pressure is reduced either by acceleration due to a reduction in the cross-sectional area or a change in height. In these adiabatic flows, cavitation is due to the local reduction in pressure whereas in boilers and heating systems it may result from a combination of increased temperature and reduced pressure. Cavitation is often detectable by the sound created by the implosive collapse of the vapour bubbles. In small-scale devices this is a harsh crackling sound whereas in very large structures, such as the spillway tunnels which carry water away from a dam, it can sound like rocks hitting the tunnel sides.

EXAMPLE 8.6
Figure E8.6 shows water flowing from a level $z' = 0$ at the bottom of a well to a level $z' = H$ at the inlet to a pump. Find the greatest depth of well H

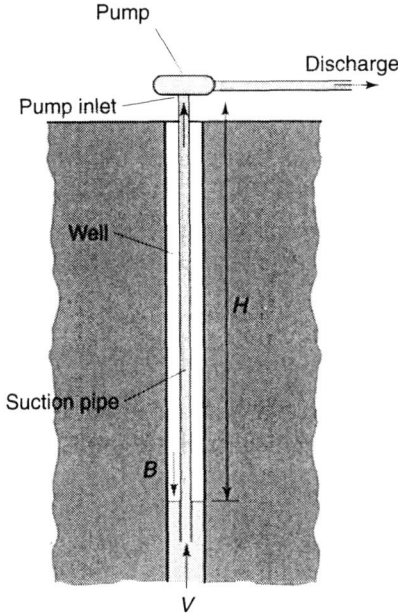

Figure E8.6

from which water can be pumped if the water surface is at atmospheric pressure B, the saturated vapour pressure of the water is p_V and the water density is ρ.

Calculate the water depth if the atmospheric pressure is 1.01 bar and the saturated vapour pressure of the water is 1.23 kPa (corresponding to 10°C).

SOLUTION
We apply Bernoulli's equation to the flow in the suction pipe, between the pump inlet and the level of the water in the well

$$p_T = p + \rho gH + \tfrac{1}{2}\rho V^2 = B + \tfrac{1}{2}\rho V^2.$$

The terms involving the velocity V cancel out and we have

$$H = \frac{B - p}{\rho g}$$

from which it is clear that H is greatest when p is as low as possible, i.e. when $p = p_V$. Thus

$$H_{\max} = \frac{B - p_V}{\rho g}.$$

For the numerical part of the problem we have

$$B = 1.01 \times 10^5 \text{ Pa}; \quad p_V = 1.23 \times 10^3 \text{ Pa}$$

Thus

$$H_{max} = \frac{(1.01 \times 10^5) - (1.23 \times 10^3)}{10^3 \times 9.81}$$

$$= 10.2 \text{ m}.$$

EXAMPLE 8.7

If the vapour pressure for the liquid in Example 8.5 is 7×10^4 Pa, what is the smallest nozzle throat diameter if the liquid is not to cavitate?

SOLUTION

If we take the throat velocity as V_t and the corresponding pressure as p_t, then we can apply Bernoulli's equation to the flow between the throat and the exit as

$$p_0 = p_t + \tfrac{1}{2} \rho V_t^2 = B + \tfrac{1}{2} \rho V^2.$$

We know already that $B = 1.01 \times 10^5$ Pa and $V = 17.5$ m/s, so that the stagnation pressure $p_0 = 2.54 \times 10^5$ Pa.

Cavitation occurs when the pressure p_t is a minimum and equal to p_V, so that

$$\tfrac{1}{2} \rho V_t^2 = p_0 - p_V = (2.54 \times 10^5) - (7 \times 10^4) = 1.84 \times 10^5 \text{ Pa}$$

and so

$$V_t = 19.2 \text{ m/s}.$$

From the continuity equation

$$\dot{m} = \rho a_t V_t$$

so that the throat cross-sectional area a_t is

$$a_t = \frac{140}{1000 \times 19.2} = 7.3 \times 10^{-3} \text{ m}^2$$

$$= \tfrac{1}{4} \pi d_t^2$$

from which the throat diameter d_t is 0.096 m or 96 mm.

The dimensionless parameter which characterises flow-induced boiling is the cavitation number Ca, defined as

$$Ca = \frac{p_{\text{REF}} - p_{\text{V}}}{\frac{1}{2}\rho V^2}$$

where p_{REF} is a reference pressure (often the ambient pressure), V is a typical velocity for the flow, p_{V} is the saturated vapour pressure and ρ is the fluid density. Cavitation within a given device occurs if the cavitation number falls below a critical value which will be dependent upon the geometry.

8.11 Summary

In this chapter we have shown how Bernoulli's equation can be applied to practical flow problems. In the case of internal flows, such as through a Venturi tube or orifice plate, we also needed the continuity equation to relate changes in cross-sectional area to changes in velocity and this restricted us to one-dimensional flows. For external flows, such as around a Pitot tube, the continuity equation was not needed and the one-dimensional restriction did not apply. For liquid flows it was shown that for sufficiently high flowspeeds the static pressure could fall below the saturated vapour pressure and lead to cavitation, i.e. internal boiling.

The student should be able to:

- identify flow problems where the application of Bernoulli's equation is appropriate
- identify flow problems where the continuity equation is also needed to solve the problem
- apply Bernoulli's equation and the continuity equation to analyse such internal flow problems as the flow through a Venturi tube or an orifice-plate flowmeter
- apply Bernoulli's equation to analyse the response of a Pitot tube or a Pitot-static tube to a fluid flow
- understand the significance of the saturated vapour pressure to liquid flow problems and calculate the flow conditions for which cavitation will occur.

8.12 Self-assessment problems

8.1 (a) A fluid of density ρ flows through a horizontal duct which contracts from a cross-section of area A_1 to a minimum (throat) area A_2. Assume one-dimensional, incompressible, frictionless flow to show that the mass flow rate $\dot{m}$ through the duct is given by

$$\dot{m} = CA_1 A_2 \sqrt{\frac{2\rho \, \Delta p}{A_1^2 - A_2^2}}$$

where Δp is the static pressure difference between the two cross-sections and C is the coefficient of discharge. Explain the significance of C.

(b) Water flows along a horizontal duct which changes from a circular pipe of diameter 100 mm to an annulus of outer diameter 100 mm and inner diameter 90 mm. Calibration tests show that the coefficient of discharge for this arrangement is 0.94. Calculate the pressure difference for a mass flow rate of 20 kg/s. Also calculate the velocity and static pressure in the throat section if the upstream stagnation pressure is 7 bar.

(Answers: 0.98 bar, 13.4 m/s, 5.99 bar)

8.2 (a) A pure liquid of density ρ flows vertically upwards through a Venturi tube which contracts from a diameter D_1 to a throat diameter D_2. If the line pressure ahead of the Venturi is p_1, show that the maximum volumetric flow rate which can be measured by the Venturi before the onset of cavitation is

$$\dot{Q} = \pi D_1^2 D_2^2 \sqrt{\frac{p_1 - \rho g S - p_V}{8\rho(D_1^4 - D_2^4)}}$$

where p_V is the vapour pressure of the liquid and S is the distance from the throat to the location where p_1 is measured. Assume frictionless flow through the Venturi tube.

(b) If the liquid in the above situation is pure water at 90°C, for which the vapour pressure is 7×10^4 N/m^2, calculate the mass flowrate corresponding to the onset of cavitation if the upstream line pressure is 2 bar, D_1 is 100 mm, D_2 is 50 mm and S is 5 m. Calculate the pressure differences for the same flowrate if the Venturi tube is operated in a horizontal water line and in a vertical water line with downflow.

(Answers: 25.8 kg/s, 0.81 bar, 0.32 bar)

8.3 (a) A liquid of density ρ and vapour pressure p_V flows through a convergent–divergent nozzle which discharges to an ambient pressure B. If the exit area A_E of the nozzle is a factor r times the throat area, show that cavitation first occurs at a flow rate $\dot{Q}$, given by

$$\dot{Q} = A_E \sqrt{\frac{2(B - p_V)}{\rho(r^2 - 1)}}.$$

Assume one-dimensional, frictionless, incompressible flow.
(b) If the throat diameter is 60 mm and the exit diameter is 90 mm, calculate the flowrate at which cavitation first occurs for a liquid of density 800 kg/m³ and a vapour pressure of 5×10^4 N/m² if the barometric pressure is 10^5 N/m². Also calculate the stagnation pressure for the flow.
(Answers: 0.0353 m³/s, 1.123 bar)

8.4 (a) A Pitot-static tube in combination with an inclined manometer is used to measure the speed V of an incompressible fluid of density ρ, as shown in Figure P8.4. If the cross-sectional area of the manometer tube is a and that of the reservoir is A, show that V is given by

$$V^2 = 2gL\left[\left(\frac{\rho_M}{\rho} - 1\right)\left(\sin\theta + \frac{a}{A}\right)\right]$$

where ρ_M is the density of the manometer fluid, θ is the inclination angle of the manometer tube, g is the acceleration due to gravity and L is the change in level of the manometer reading (i.e. the level change measured along the tube).
(b) A Pitot-static tube is used to measure the flowspeed of a liquid of density 1200 kg/m³. The manometer fluid is mercury for which the relative density is 13.6. The internal diameter of the manometer tube is 5 mm and that of the reservoir 100 mm. Calculate the inclination angle θ if the manometer reading L is to be 500 mm for a flow with stagnation pressure 1.16 bar and static pressure 1.01 bar.

8.5 (a) A jet of liquid of density ρ, surrounded by air, flows vertically downwards from a nozzle of area A_0. If the stagnation pressure of the jet at exit from the nozzle is p_0 and the surrounding air is at pressure B, show

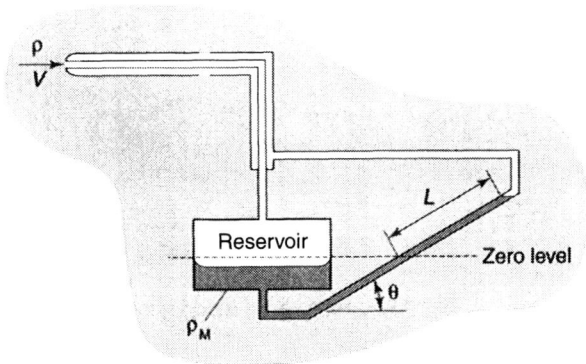

Figure P8.4

that the cross-sectional area of the jet A changes with vertical distance below the nozzle exit z according to

$$\left(\frac{A_0}{A}\right)^2 = 1 + \frac{\rho g z}{p_0 - B}$$

where g is the acceleration due to gravity. Assume the flow is one dimensional and frictionless and regard the air density as negligible.

(b) If the nozzle area A_0 is 5×10^{-4} m^2 and the stagnation pressure p_0 is 2 bar, determine the jet velocity a distance 1000 m below the nozzle if the liquid is aviation fuel of density 700 kg/m^3 and the ambient pressure B is 0.5 bar.

(Answer: 141.6 m/s)

8.6 (a) The arrangement shown in Figure P8.6 is used to inject liquid detergent into the water flowing through a fire hose in order to create foam. The cross-sectional area of the contraction at section ① is A_C and the cross-sectional area of the outlet nozzle is A_E. The stagnation pressure of the water flow is p_0, the ambient pressure to which the nozzle discharges is B, and the contraction height above the surface of the detergent pool is H. Find a relationship between A_C, A_E, p_0, B, ρ, g and H for which detergent just rises to the top of the vertical tube. Assume one-dimensional, incompressible, frictionless flow for the water and hydrostatic conditions, with gravitational acceleration g, for the detergent which has the same density as water.

(b) If the outlet nozzle in Figure P8.6 has diameter 70 mm and the internal diameter of the contraction is 60 mm, calculate the stagnation pressure for a water mass flowrate of 55 kg/s if the ambient pressure is 1.02 bar. Calculate also the static pressure at the location of the contraction. What

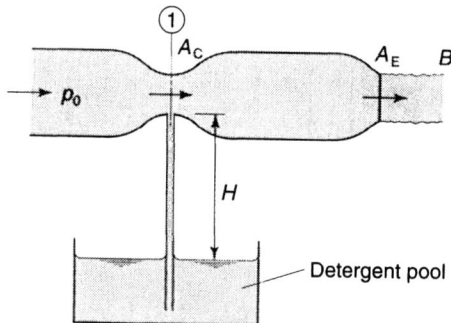

Figure P8.6

is the maximum vertical height difference between the contraction and the
surface of the detergent pool if the detergent is to rise to the top of the
vertical tube? If the vapour pressure of water is taken as 2.3 kPa, calculate
the water mass flowrate at which cavitation occurs.
(Answers: 2.04 bar, 0.15 bar, 8.9 m, 58.9 kg/s)

CHAPTER 9

Linear momentum equation and hydrodynamic forces

This brief but important chapter is concerned with fluid flow through a duct which changes in cross-sectional area and/or direction. Force must be exerted on the fluid to produce the changes in fluid momentum which are a consequence of such geometric changes. What is of interest from an engineering point of view is the external reaction force which has to be applied to counteract the force exerted by the fluid on the interior surface of the duct. We use Newton's second law of motion to derive the linear momentum* equation for a flowing fluid. We then identify the separate contributions to the net force acting on the fluid due to the fluid pressure at inlet and outlet to the duct, and the force exerted on the fluid by the duct walls. We exclude from the analysis any body forces, including the weight of the fluid. The analysis is completed by applying the principle of static equilibrium to equate the internal and external forces acting on the duct. Emphasis is given to the vector nature of force and momentum flowrate.

9.1 Problem under consideration

In this chapter we consider the flow of fluid through a duct, such as is illustrated in Figure 9.1, which may have a varying cross-section of any shape and the centreline of which may be straight or curved. The cross-sectional area of the duct at any location s along its centreline is denoted by $A(s)$. The word duct is used to mean any passage or channel through which there is fluid flow and includes, for example, pipes, pipe bends, nozzles, Venturi tubes, diffusers, engine intakes and exhausts and rocket engines. We retain the assumption of steady, one-dimensional flow, but allow the interaction between the flowing fluid and the duct walls to involve not only pressure $p(s)$ but also shear stress $\tau(s)$ due to the fluid viscosity, i.e. we no longer assume that the flow is frictionless. The restriction to constant density is also dropped for the basic analysis.

---

* We are concerned with linear momentum because, for the flows under consideration, the effects of rotation of the fluid about an axis can be neglected. If we were to consider the internal flow within, for example, the blading of an axial-flow turbine or the impeller of a centrifugal compressor, the effects of fluid rotation would be important and it would be essential to consider the torque acting on the fluid and its angular momentum.

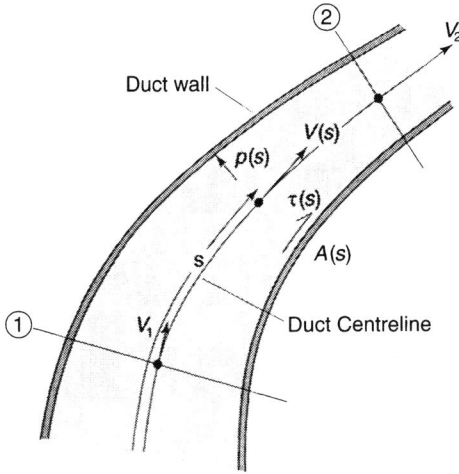

Figure 9.1 *Fluid flow through a duct*

As indicated in Figure 9.1, we shall apply our analysis to a segment of the duct with an inlet (section ①) and an outlet (section ②). In many instances, sections ① and ② will correspond to an actual inlet or outlet, for example the intake to a jet engine or a nozzle exit. In other situations – for example, for flow through a complex duct system – an essential part of the analysis is to identify an appropriate duct segment for analysis. We shall illustrate this point in some of the examples in Chapter 10. The volume between the inlet and outlet defined by the *wetted (i.e. interior) surface* of the duct segment is referred to as a *control volume*.

As we have seen in Chapter 6, changes in the cross-sectional area of a duct result in changes in the velocity of a fluid flowing through the duct. As we shall show in the section which follows that if, because the fluid has mass, its velocity changes, then so must a quantity we call the *momentum flowrate* $\dot{\mathbf{M}}^*$ and this requires that a force is applied to the fluid. A force applied to the fluid is also required to change the flow direction.

The forces acting on the fluid within the control volume are shown in Figure 9.2(a). The net force **F** arises from the pressures at inlet and outlet, p_1 and p_2, and from the pressure and shear stress distributed over the wetted surface of the duct, $p(s)$ and $\tau(s)$. The net force due to $p(s)$ and $\tau(s)$ we refer to as the *fluid–structure interaction force* **S**. According to Newton's third law of motion, the fluid in the control volume must exert on the wetted surface of the duct a force equal in magnitude but opposite in direction to **S**. As indicated in Figure 9.2(b), for the duct segment to be in *static equilibrium*, this force must be balanced by the *external*

---

* Vector quantities (i.e. those having both a magnitude and a direction) are shown in bold-face type.

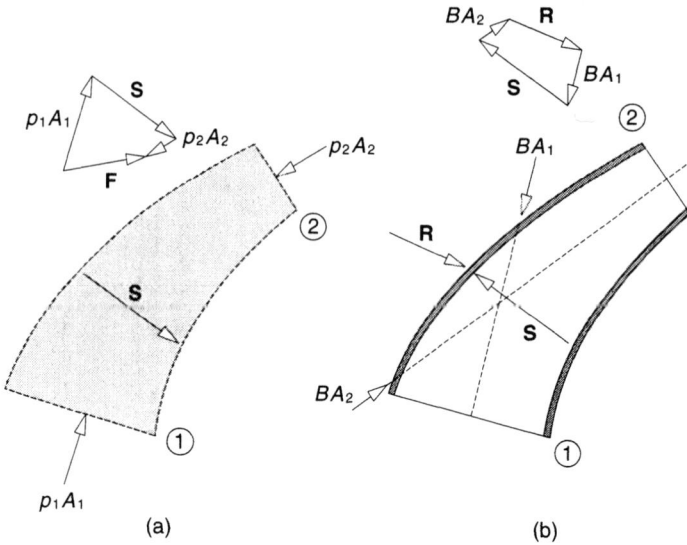

Figure 9.2 *Forces acting on (a) fluid control volume and (b) duct segment*

forces acting on the duct, usually made up of a force due to the *external pressure B* distributed over its outer surface and an *applied restraining (or reaction) force* **R**. Polygons illustrating the vector addition of all of these forces are given in Figure 9.2.

The aim of this chapter is to relate the forces **F**, **S** and **R**, the external pressure *B*, the pressures at inlet and outlet p_1 and p_2, the fluid mass flowrate $\dot{m}$, the fluid density ρ, and the cross-sectional areas at inlet and outlet A_1 and A_2. As will become apparent in Chapter 10, the results of this chapter can be applied directly even if there is more than one inlet or outlet. Also, although for simplicity we restrict attention to flows for which the velocity and all other vector quantities have components only in two perpendicular directions, x and y, the analysis is easily extended to include the third direction.

9.2 Basic linear momentum equation

In Chapter 7 we applied Newton's second law of motion to a fluid slice flowing through a streamtube. In much the same way, as shown in Figure 9.3, we now apply Newton's second law of motion to a fluid slice flowing through the control volume. As should already be evident, it is crucial that we allow for the fact that *force*, *acceleration*, *velocity* and *momentum* are all vector quantities, which we do by considering separately the x- and y-directions.

We consider first the x-direction. If the mass of the fluid slice is δm then from

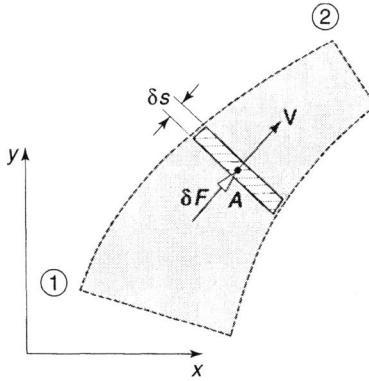

Figure 9.3 *Fluid slice flowing through control volume*

Newton's second law of motion we have

$$\delta F_x = \delta m \, a_x \qquad (9.1)$$

where δF_x is the component of force acting on the fluid slice in the x-direction and a_x is the corresponding acceleration of the slice in the x-direction. Since the flow is steady, the acceleration is a consequence of changes in the velocity of the fluid due to changes in the cross-sectional area of the duct, i.e. in a similar way to section 7.2 we may write

$$a_x = \frac{dV_x}{dt} = \frac{dV_x}{ds}\frac{ds}{dt} = V\frac{dV_x}{ds}$$

where s is the distance measured along the duct centreline.

Substituting the final expression for a_x into equation (9.1) gives

$$\delta F_x = \delta m \, V \frac{dV_x}{ds}$$

$$= \rho A \, \delta s \, V \frac{dV_x}{ds}.$$

where we have replaced the mass of the slice by $\rho A \, \delta s$.

From the continuity equation (6.1) we have

$$\dot{m} = \rho A V$$

so that our equation can be written as

$$\delta F_x = \dot{m} \frac{dV_x}{ds} \, \delta s$$

or, in the limit of an infinitesimally thin fluid slice $\delta s \rightarrow 0$

$$\frac{dF_x}{ds} = \dot{m}\frac{dV_x}{ds}$$

or

$$dF_x = \dot{m}\,dV_x = d\dot{M}_x \tag{9.2}$$

where

$$\dot{M}_x = \dot{m}V_x$$

is the x-component of the *momentum flowrate* (or *momentum flux*) of the fluid flowing through the duct. For a solid mass m moving with velocity V, the quantity mV is called the momentum. In a similar way, for a fluid stream, the quantity $\dot{m}V$ is called the momentum flowrate or flux. What equation (9.2) shows is that the x-component of the fluid velocity and the corresponding momentum flowrate will increase if the x-component of force acting on the fluid slice is positive.

Integration of equation (9.2) along the duct centreline between locations 1 and 2 produces the important result

$$\boxed{F_x = \dot{m}(V_{2_x} - V_{1_x}) = \dot{M}_{2_x} - \dot{M}_{1_x}} \tag{9.3}$$

where F_x is the x-component of the net force acting on the fluid within the control volume, V_{2_x} is the x-component of the fluid velocity leaving the control volume and V_{1_x} is the x-component of the fluid velocity entering the control volume. In equation (9.3), $\dot{M}_{2_x}$ is the x-component of the momentum flux at exit from the control volume and $\dot{M}_{1_x}$ is the x-component of the momentum flux at inlet to the control volume.

It should be apparent that the corresponding result to equation (9.3) for the y-direction is

$$\boxed{F_y = \dot{m}(V_{2_y} - V_{1_y}) = \dot{M}_{2_y} - \dot{M}_{1_y}} \tag{9.4}$$

According to equations (9.3) and (9.4), the *net force acting on the fluid in the control volume in a given direction is equal to the change in the momentum flowrate (i.e. the rate of change of fluid momentum) in the same direction.* The triangle included in Figure 9.4 illustrates that the net force **F** is equal to the vector difference between the momentum flowrates out of and into the control volume.

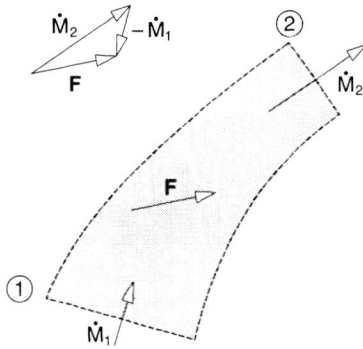

Figure 9.4 *Net force and momentum flowrates for control volume*

EXAMPLE 9.1

A fluid of density ρ flows with a mass flowrate $\dot{m}$ through a pipe which turns through 90° and at the same time halves in cross-sectional area. If the initial cross-sectional area is A, find the net force exerted on the fluid within the pipe bend.

SOLUTION

The problem under consideration is illustrated in Figure E9.1(a) and the corresponding fluid control volume in Figure E9.1(b). The approach flow direction is taken as x and the outflow direction as y.

The momentum equation for the x-direction gives

$$F_x = 0 - \dot{m}V_1 = -\frac{\dot{m}^2}{\rho A}$$

where we have made use of the continuity equation

$$\dot{m} = \rho A V_1.$$

For the y-direction we have

$$F_y = \dot{m}V_2 - 0 = \frac{\dot{m}^2}{\rho A_2} = \frac{2\dot{m}^2}{\rho A}$$

where we have again used the continuity equation and also the area relation

$$A_2 = \tfrac{1}{2} A.$$

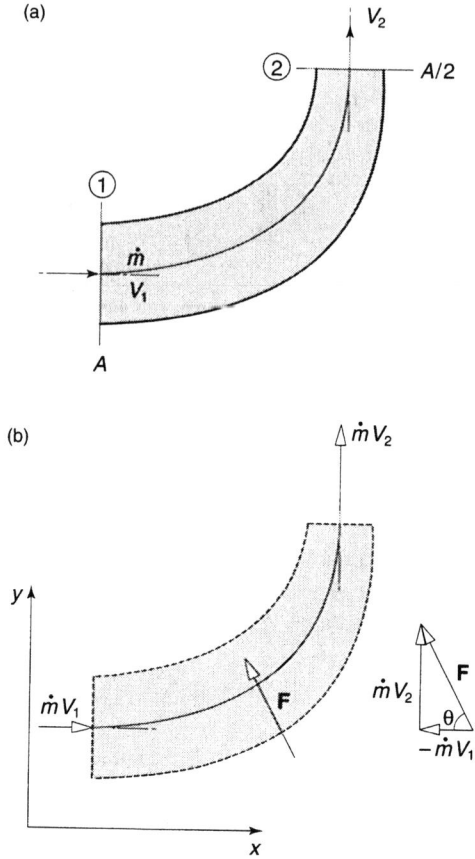

(a)

(b)

Figure E9.1

The magnitude of the net force **F** acting on the fluid is then

$$F = \sqrt{(F_x^2 + F_y^2)} = \frac{\sqrt{5} \times \dot{m}^2}{\rho A}$$

and this force acts at an angle θ given by

$$\tan \theta = \frac{F_y}{F_x} = -2 \quad \text{i.e.} \quad \theta = -63.4°.$$

Comment: It is essential to draw a diagram showing the general arrangement of the flow situation under consideration, including symbols for any specified quantities, reference axes, etc. To further aid in the solution, draw a second diagram showing the control volume, again including all relevant information.

Note that in the equation for F_x, the outflow momentum in the x-direction is zero and so F_x itself is negative, i.e. opposed to the direction of the approach flow. For the y-direction, on the other hand, the outflow momentum is positive but the inflow momentum is zero and F_y is positive. This is entirely what we should expect since the action of turning the flow through 90° requires that its initial momentum in one direction is reduced to zero while the momentum in the perpendicular direction is increased from zero to a value $\dot{m}V_2$. In this case $V_2 = 2V_1$ so that the magnitude of the momentum outflow $\dot{m}V_2$ is double that of the momentum inflow $\dot{m}V_1$.

9.3 Net force on fluid in control volume

As pointed out in section 9.1 and illustrated in Figure 9.2(a), the net force **F** includes the pressure forces $p_1 A_1$ and $p_2 A_2$, and the *fluid–structure interaction force* **S**.

For the x-direction we have

$$F_x = (p_1 A_1)_x - (p_2 A_2)_x + S_x = \dot{m}(V_{2_x} - V_{1_x}) \qquad (9.5)$$

and for the y-direction

$$F_y = (p_1 A_1)_y - (p_2 A_2)_y + S_y = \dot{m}(V_{2_y} - V_{1_y}) \qquad (9.6)$$

where the subscripts x and y again indicate the components of the vector quantities **F**, **pA**, **S** and **V** in the x- and y-directions, respectively.

In order to proceed further, the pressures p_1 and p_2 have to be specified or calculated. If, as is often the case, the flow can be regarded as frictionless, Bernoulli's equation can be used to relate p_1 and p_2. If the flow discharges to atmosphere at pressure B, then $p_2 = B$. In other situations it may be necessary to calculate p_1 and p_2 from other information, or the values may be available from measurements.

EXAMPLE 9.2
In the pipe bend situation considered in Example 9.1, the flow discharges to atmospheric pressure B and can be assumed frictionless and incompressible. Gravitational effects may also be ignored. Find the force exerted on the fluid by the internal surface of the pipe bend.

SOLUTION
Figure E9.2 again shows the fluid control volume, but now includes the pressures p_1 and p_2 $(=B)$ at inlet and outlet, and also the force exerted on

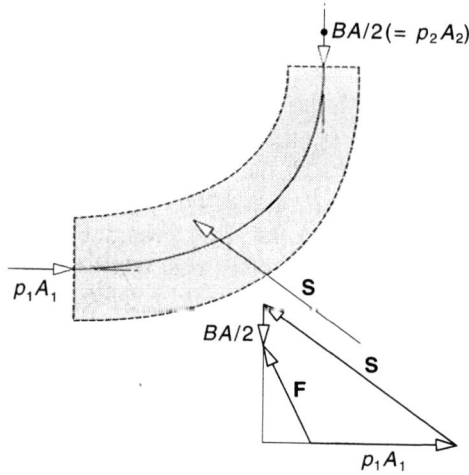

Figure E9.2

the fluid in the control volume by the wetted surface of the pipe, i.e. the fluid–structure interaction force **S**.
For the x-direction we have

$$F_x = p_1 A - S_x$$

and for the y-direction

$$F_y = -p_2 A_2 + S_y = -B\frac{A}{2} + S_y$$

wherein we have made use of the fact that the fluid discharges to the atmospheric pressure, i.e. $p_2 = B$, and also that the outlet area is half the inlet area, i.e.

$$A_2 = \tfrac{1}{2} A.$$

Since the flow can be assumed frictionless, we can introduce Bernoulli's equation to evaluate the unspecified inlet pressure p_1, i.e.

$$p_1 + \tfrac{1}{2}\rho V_1^2 = B + \tfrac{1}{2}\rho V_2^2.$$

As before, the velocities V_1 and V_2 can be determined from the continuity equation as

$$V_1 = \frac{\dot{m}}{\rho A} \quad \text{and} \quad V_2 = \frac{2\dot{m}}{\rho A}$$

so that

$$p_1 = B + \frac{3}{2\rho}\left(\frac{\dot{m}}{A}\right)^2 \quad \text{and} \quad F_x = BA + \frac{3\dot{m}^2}{2\rho A} - S_x.$$

From Example 9.1 we have

$$F_x = -\frac{\dot{m}^2}{\rho A} \quad \text{and} \quad F_y = \frac{2\dot{m}^2}{\rho A}$$

which we can now combine with the equations for F_x and F_y obtained here to give

$$S_x = BA + \tfrac{5}{2}\frac{\dot{m}^2}{\rho A} \quad \text{and} \quad S_y = \tfrac{1}{2}BA + \frac{2\dot{m}^2}{\rho A}.$$

The magnitude and direction of **S** then follow from

$$S = \sqrt{(S_x^2 + S_y^2)} \quad \text{and} \quad \tan\phi = \frac{S_y}{S_x}.$$

9.4 Hydrodynamic reaction force

So far in this chapter we have been concerned with the forces exerted on the fluid within the control volume and with the fluid–structure interaction force **S**. However, in the majority of engineering applications it is the *reaction force* **R** required to restrain the structure which is of principal interest. At first sight it might appear that the reaction force would be equal in magnitude to the force exerted on the duct walls, i.e. the fluid–structure interaction force **S**. That this is not so should be apparent from the free-body diagram shown in Figure 9.2(b) which shows that the external pressure B must also be taken into account.   For the duct to be in static equilibrium we require

$$R_x + (BA_2)_x - (BA_1)_x - S_x = 0 \tag{9.7}$$

and

$$R_y + (BA_2)_y - (BA_1)_y - S_y = 0 \tag{9.8}$$

where R_x and R_y are the x- and y-components of the reaction force **R**.

The form of the terms involving the external pressure B can be explained as follows. As we showed in section 5.1, the resultant force acting on a body completely subjected to uniform pressure over its entire outer surface is zero. In the present case, since there are openings in the outer surface at the inlet and outlet, unbalanced forces corresponding to BA_1 and BA_2 must arise on the opposite side of the structure, as shown in Figure 9.2(b). It is the x- and y-components of these two forces which appear in equations (9.7) and (9.8).

If we now combine equations (9.5) and (9.7) we find

$$R_x = \dot{m}(V_{2_x} - V_{1_x}) - [(p_1 - B)A_1]_x + [(p_2 - B)A_2]_x \qquad (9.9)$$

and from (9.6) and (9.8) we have

$$R_y = \dot{m}(V_{2_y} - V_{1_y}) - [(p_1 - B)A_1]_y + [(p_2 - B)A_2]_y \qquad (9.9)$$

Equations (9.9) and (9.10) reveal that for the calculation of the *reaction force* it is the so-called *gauge pressures* $p_1 - B$ *and* $p_2 - B$ which are important rather than the absolute pressure levels p_1 and p_2. In fact, as the following example illustrates, the reaction force $\mathbf{R}$ for a given duct frequently depends only upon the flowrate $\dot{m}$ and the fluid density ρ and so is called the *hydrodynamic reaction force*.

EXAMPLE 9.3
Calculate the external reaction force needed to hold in place the pipe bend of Example 9.1.

SOLUTION
Once again it is valuable to draw a diagram to aid in the solution, this time showing the internal and external forces acting on the pipe bend.

For the bend to be in static equilibrium we require

$$-R_x - BA + S_x = 0 \quad \text{and} \quad R_y + \tfrac{1}{2}BA - S_y = 0$$

so that

$$R_x = S_x - BA \quad \text{and} \quad R_y = S_y - \tfrac{1}{2}BA$$

From Example 9.2 we thus find

$$R_x = \frac{5\dot{m}^2}{2\rho A} \quad \text{and} \quad R_y = \frac{2\dot{m}^2}{\rho A}$$

so that the magnitude of the external reaction force is

$$R = \sqrt{(R_x^2 + R_y^2)} = \frac{\sqrt{41} \times \dot{m}^2}{2\rho A}$$

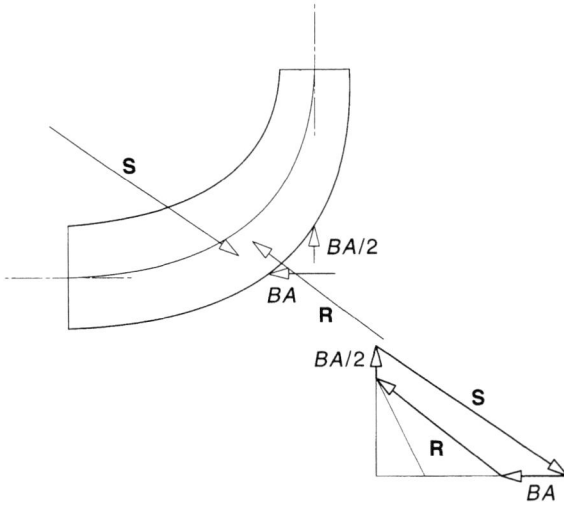

Figure E9.3

and the direction of the external reaction force is given by

$$\tan \xi = \frac{R_y}{R_x} = \tfrac{4}{5} \quad \text{i.e.} \quad \xi = 38.7°.$$

Comment: This example underlines the point made earlier that the external restraining force differs from the force exerted on the wetted surface of the pipe bend by the fluid flowing through it as a consequence of the pressure of the surroundings, and the final result is independent of the external pressure.

EXAMPLE 9.4
A pipe bend such as that shown in Figure E9.1 has an upstream diameter of 1 m. With water flowing through the bend, the external force required to hold it in place is 1200 N. Calculate the volume flowrate of the water and the pressure drop through the bend.

SOLUTION

$$D_1 = 1 \text{ m}; \ R = 1200 \text{ N}; \ \rho = 1000 \text{ kg/m}^3$$

We calculate first the cross-sectional area A:

$$A = \tfrac{1}{4} \pi D_1^2 = 0.785 \text{ m}^2.$$

From Example E9.3 we have

$$R = \frac{\sqrt{41} \times \dot{m}^2}{2\rho A}$$

which gives

$$\dot{m} = \sqrt{\frac{1200 \times 2 \times 1000 \times 0.785}{\sqrt{41}}} = 543 \ \text{kg/s}$$

and so the volumetric flowrate

$$\dot{Q} = \frac{\dot{m}}{\rho} = 0.543 \ \text{m}^3/\text{s}.$$

To calculate the pressure drop $\Delta p = p_1 - p_2$ through the bend we use Bernoulli's equation, i.e.

$$p_1 + \tfrac{1}{2}\rho V_1^2 = p_2 + \tfrac{1}{2}\rho V_2^2.$$

Since the cross-sectional area halves, according to the continuity equation the fluid velocity must double, so that

$$\Delta p = p_1 - p_2 = \tfrac{3}{2}\rho V_1^2$$

$$= 1.5 \times 1000 \times \left(\frac{0.543}{0.785}\right)^2$$

$$= 716 \ \text{Pa} \quad \text{or} \quad 7.16 \times 10^{-3} \ \text{bar}.$$

9.5 Summary

The overall objective of this chapter was to show how we can calculate the external force which must be applied to a duct to counteract the hydrodynamic forces generated by a fluid flowing through the duct. After introducing the concept of a fluid control volume, we showed that the analysis involves three separate considerations. We started by applying Newton's second law of motion to fluid flow through a duct of arbitrary shape. The outcome was the linear momentum equation which shows that the change in momentum flowrate of the fluid is equal to the net force exerted on the fluid. The second step was to identify the individual forces which contribute to the net force: the pressure forces at inlet and outlet and the forces exerted on the fluid due to the pressure and shear stress distributed over the wetted interior surface of the duct. Finally we used the condition of static equilibrium for the duct to relate the external restraining force to the force exerted by the flowing fluid on the wetted interior surface, which we termed the fluid–structure interaction force. The vector nature of force and momentum flowrate was accounted for by considering components in given directions. The student should be able to

- explain the concept of a control volume
- apply the linear momentum equation

$$\boxed{F_x = \dot{m}(V_{2_x} - V_{1_x})} \text{ and } \boxed{F_y = \dot{m}(V_{2_y} - V_{1_y})}$$

to fluid flow through a specified control volume
- write down expressions for the separate forces which contribute to the net force acting on the fluid flowing through a control volume, i.e. the pressure forces acting on the fluid at inlet and outlet to the control volume and the fluid–structure interaction force
- use the condition of static equilibrium for the duct to write down equations relating the external reaction force, the fluid–structure interaction force and the forces due to the pressure of the surroundings.

9.6 Self-assessment problems

9.1 A pipe of cross-sectional area A turns through 90°. A fluid of density ρ with static pressure equal to the external pressure flows through the pipe with mass flowrate $\dot{m}$. Draw two diagrams, one of the fluid control volume including the forces and momentum flowrates and the other a free-body diagram showing the forces acting on the bend. Show that the net force required to hold the bend in place is $(\sqrt{2} \times \dot{m}^2)/\rho A$. Assume that the flow is one dimensional and friction free.

9.2 Fluid of density ρ flows through a turbomachine at a mass flowrate $\dot{m}$. If the inlet area is A and the outlet area is fA, where f is a numerical factor, derive a formula for the net force F on the fluid within the turbomachine in terms of $\dot{m}, \rho, f$ and A if the angle between the entry and outlet directions is 60°.

9.3 A fluid of density ρ flows at pressure p and velocity V through a pipe of cross-sectional area A. If the pipe turns through 90°, find the force exerted on the fluid by the pipe wall. The flow may be assumed to be one dimensional and frictionless, and gravity effects may be neglected.

9.4 (a) A liquid of density ρ flows through a circular pipe with diameter D at a mass flowrate $\dot{m}$. The pipe turns through 90° in the horizontal plane and also reduces in diameter by 50%. The hydrodynamic loads exerted on the bend have to be supported externally. Assume one-dimensional frictionless flow to show that the components of the external reactions R_x and R_y are

$$R_x = \frac{\pi}{4}(p-B)D^2 + \frac{4}{\pi\rho}\left(\frac{\dot{m}}{D}\right)^2$$

and

$$R_y = \frac{\pi}{16}(p-B)D^2 + \frac{17}{2\pi\rho}\left(\frac{\dot{m}}{D}\right)^2$$

where R_x is in the opposite direction to the approach flow and R_y is in the direction of the leaving flow. B is the external pressure and p is the static pressure of the approach flow.

(b) If the flow described in part (a) discharges at the exit from the bend to an external pressure B of 1 bar, the liquid has a density of 2×10^3 kg/m³, the flowrate is 100 kg/s and the upstream pipe diameter is 150 mm, calculate the resultant force required to support the bend and the direction of its line of action.

(Answers: 2658 N, 25.2°)

9.5 (a) Water flows through a pipe which turns through 180° in a horizontal plane and at the same time doubles in diameter. If the flow discharges at atmospheric pressure, show that the magnitude of the force R needed to hold the pipe in place is given by

$$R = \frac{25\dot{m}^2}{32\rho A}$$

where $\dot{m}$ is the mass flowrate of the water, ρ the water density and A the cross–sectional area of the pipe before the bend. Assume incompressible, one-dimensional, frictionless flow and ignore gravity effects.

(b) The external force required to hold a pipe bend in place is 1500 N. Calculate the volume flowrate of water if the pipe diameters are 0.5 m upstream of the bend and 1.0 m downstream. Also calculate the pressure change through the bend.

(Answers: 0.614 m³/s, 0.046 bar)

CHAPTER 10

Engineering applications of the linear momentum equation

In this, the final, chapter we use a number of specific problems to show how the analysis of Chapter 9 can be applied to real engineering situations. Each problem, together with the numerical example which follows it, can be thought of as a case study. It is important not to think of the numerical examples in isolation, to be solved simply by substituting numbers into formulae. In an examination, the student would be expected to carry out the theoretical analysis first starting from the basic equations of momentum, static equilibrium, continuity and, if appropriate, Bernoulli.

We consider flow through a convergent nozzle, a rocket engine, a turbojet engine, a turbofan engine, a sudden enlargement and a jet pump – all examples where the generation of hydrodynamic force and changes in pressure, velocity and momentum flowrate are entirely due to changes in cross-sectional area. The remaining problems involve the additional complication of a change in flow direction, as in the internal flow through a pipe bend, a pipe junction and a set of curved guidevanes; and finally two situations where a free jet is deflected by impingement on a fixed or moving object.

10.1 Force required to restrain a convergent nozzle

The conventional technique for dousing a major fire relies on a convergent nozzle to create a high-speed jet of water which is directed towards the fire. Water at high pressure (typically 3 bar) flows at low speed through a long hose (typically 80 mm internal diameter) and is discharged into the surroundings at atmospheric pressure and high speed through a nozzle typically 50 mm in diameter. As we shall see, the nozzle generates a considerable reaction force which has to be resisted by the firefighter.

As shown in Figure 10.1(a), we consider a convergent nozzle, of exit area A_2, connected to a hose of cross-sectional area A_1, through which an incompressible fluid of density ρ flows at a mass flowrate $\dot{m}$. The fluid discharges to the surroundings which are at a pressure B. Our aim is to determine the magnitude and direction of the restraining force required to hold the nozzle in place.

In this instance there is only one choice for the fluid control volume, as shown in Figure 10.1(b). Since there is no change in flow direction, we need consider

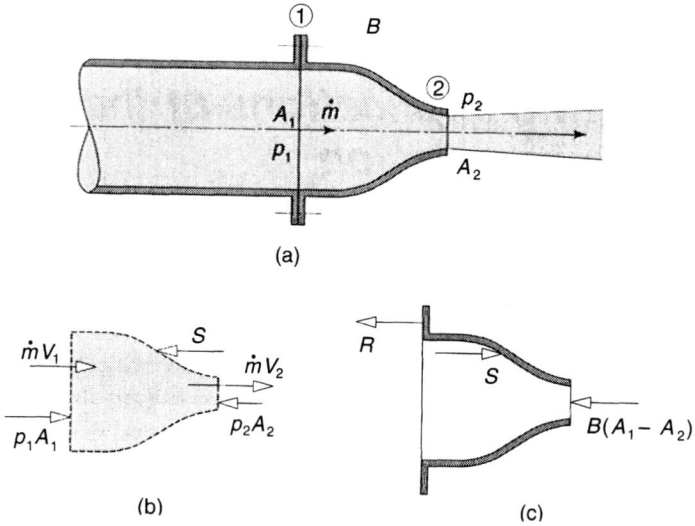

Figure 10.1 (a) Nozzle arrangement; (b) fluid control volume; (c) forces acting on nozzle

only the axial direction. According to the linear momentum equation, the net force F. exerted on the fluid in the control volume is equal to the difference in momentum flowrate between outlet and inlet, i.e.

$$F = \dot{M}_2 - \dot{M}_1 = \dot{m}(V_2 - V_1).$$

The flow velocities V_1 and V_2 can be found from the continuity equation as

$$V_1 = \frac{\dot{m}}{\rho A_1} \quad \text{and} \quad V_2 = \frac{\dot{m}}{\rho A_2}$$

so that

$$F = \frac{\dot{m}^2}{\rho}\left(\frac{1}{A_2} - \frac{1}{A_1}\right)$$

As indicated in Figure 10.1(b), there are three contributions to the force F: the pressure forces $p_1 A_1$ and $p_2 A_2$, and the force exerted directly on the fluid by the internal surface of the nozzle S, i.e.

$$F = p_1 A_1 - p_2 A_2 - S = \frac{\dot{m}^2}{\rho}\left(\frac{1}{A_2} - \frac{1}{A_1}\right). \tag{10.1}$$

Our aim is to calculate the external reaction force R required to restrain the nozzle against the forces imposed on it. As shown in Figure 10.1(c), these forces comprise the internal force exerted directly on the fluid by the nozzle S (equal in magnitude but opposite in direction to the force exerted on the fluid) and the forces due to the external pressure B.

From the condition for static equilibrium we have

$$R - S + BA_1 - BA_2 = 0$$

which we can combine with equation (10.1) to eliminate S to find

$$R = \frac{-\dot{m}^2}{\rho}\left(\frac{1}{A_2} - \frac{1}{A_1}\right) + (p_1 - B)A_1 - (p_2 - B)A_2.$$

Since the nozzle discharges to the surroundings at ambient pressure B, we can take $p_2 = B$ to give

$$R = \frac{-\dot{m}^2}{\rho}\left(\frac{1}{A_2} - \frac{1}{A_1}\right) + (p_1 - B)A_1. \tag{10.2}$$

Equation (10.2) is as far as we can take our calculation without further information about the upstream pressure p_1. If the flow can be regarded as frictionless, p_1 can be calculated from Bernoulli's equation, i.e.

$$p_0 = p_1 + \tfrac{1}{2}\rho V_1^2 = p_2 + \tfrac{1}{2}\rho V_2^2$$

where p_0 is the stagnation pressure of the flow. As before, we can use the continuity equation to obtain V_1 and V_2, so that Bernoulli's equation can be written as

$$p_1 - B = \frac{\dot{m}^2}{2\rho}\left(\frac{1}{A_2^2} - \frac{1}{A_1^2}\right)$$

which we can substitute into equation (10.2) to obtain, after some simplification,

$$R = \frac{\dot{m}^2 A_1}{2\rho}\left(\frac{1}{A_2} - \frac{1}{A_1}\right)^2. \tag{10.3}$$

This final result shows a number of features which are typical of fluid flow calculations. First, as must be the case, in the absence of flow ($\dot{m} = 0$) the reaction force $\mathbf{R}$ is zero. Second, the relationship between the reaction force and the flowrate is quadratic. Although this result has come about from an analysis in which frictional effects have been excluded, it turns out that for many practical flow situations where the flow is highly turbulent and there are substantial frictional losses, hydrodynamic forces are proportional to the square of the flowrate or velocity.

EXAMPLE 10.1

Water flows through a hose with an internal diameter of 80 mm and discharges to the atmosphere through a convergent nozzle with an exit diameter of 60 mm. If the mass flowrate of the water is 50 kg/s, calculate the force required to restrain the nozzle. If the ambient pressure is 1.03 bar, calculate the static pressure of the flow in the hose and also the stagnation pressure of the flow which may be assumed to be frictionless.

SOLUTION

$D_1 = 0.08$ m; $D_2 = 0.06$ m; $\dot{m} = 50$ kg/s; $p_2 = B = 1.03 \times 10^5$ Pa; $\rho = 1000$ kg/m^3.

We calculate first the areas A_1 and A_2:

$$A_1 = \tfrac{1}{4}\,\pi D_1^2 = 5.03 \times 10^{-3}\ \text{m}^2$$
$$A_2 = \tfrac{1}{4}\,\pi D_2^2 = 2.83 \times 10^{-3}\ \text{m}^2$$

Substitution in equation (10.3) then gives

$$R = \frac{0.5 \times 50^2 \times 5.03 \times 10^{-3}}{1000}\left(\frac{1}{2.83} - \frac{1}{5.03}\right)^2 \times 10^6$$

$$= 150.4\ \text{N}.$$

Since the flow is frictionless, we can use Bernoulli's equation to relate the water pressures in the hose and at the nozzle exit, i.e.

$$p_0 = p_1 + \tfrac{1}{2}\,\rho V_1^2 = B + \tfrac{1}{2}\,\rho V_2^2.$$

The velocities V_1 and V_2 are calculated from the continuity equation

$$\dot{m} = \rho A_1 V_1 = \rho A_2 V_2$$

from which

$$V_1 = \frac{50}{1000 \times 5.03 \times 10^{-3}} = 9.95\ \text{m/s}$$

and

$$V_2 = \frac{50}{1000 \times 2.83 \times 10^{-3}} = 17.68 \text{ m/s}.$$

The stagnation pressure p_0 is then

$$p_0 = B + \tfrac{1}{2}\rho V_2^2 = 1.03 \times 10^5 + 0.5 \times 1000 \times 17.68^2$$

$$= 2.59 \times 10^5 \text{ Pa} \quad \text{or} \quad 2.59 \text{ bar}$$

and the upstream pressure p_1 is

$$p_1 = p_0 - \tfrac{1}{2}\rho V_1^2 = 2.59 \times 10^5 - 0.5 \times 1000 \times 9.95^2$$

$$= 2.10 \times 10^5 \text{ Pa} \quad \text{or} \quad 2.10 \text{ bar}.$$

10.2 Rocket-engine thrust

Figure 1.12 shows the general configuration for a liquid-propellant rocket engine. A typical propellant would be liquid hydrogen as the fuel with liquid oxygen as the oxidiser. As the propellant burns, the gas (at high temperature and pressure) which is produced is exhausted through a divergent exhaust nozzle into the surroundings as a high-velocity jet. Although the combustion chamber-exhaust nozzle arrangement resembles a Venturi tube, it functions very differently because the gas flow through it is highly compressible. The gas flow is subsonic in the combustion chamber itself, accelerates to sonic conditions (i.e. Mach number of 1) at the nozzle throat and becomes supersonic in the diffuser. Unlike the situation for subsonic flow, the pressure of the jet flow at exit from the diffuser is usually higher than that of the surrounding atmosphere which, at high altitude (above 30 000 m), is almost zero. As we shall now show, the thrust produced by a rocket engine is equal to the momentum flux of the exhaust jet plus the pressure force due to the difference between the exhaust pressure and the ambient pressure.

As always, it is convenient to consider gas flow relative to the rocket engine, as would be the case if it were on a test bed. As was the case for the convergent nozzle of section 10.1, the interior surface of the rocket engine is a suitable choice to define a fluid control volume, as shown in Figure 10.2(a). The momentum fluxes associated with the inflow of fuel and oxidant will be negligible and the momentum equation applied to the fluid control volume is simply

$$F = \dot{m}V_E \tag{10.4}$$

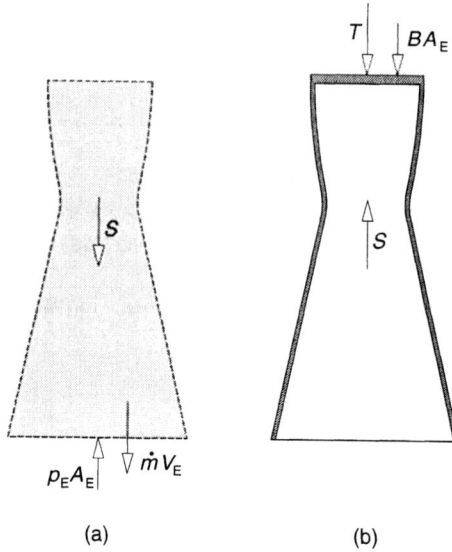

Figure 10.2 *Liquid-propellant rocket engine: (a) fluid control volume; (b) forces acting on rocket engine*

where $\dot{m}$ is the mass flowrate of the exhaust gas, which for steady flow must equal the combined mass flowrate of propellant and oxidant. In the case of a solid-propellant rocket, the only difference to the liquid-propellant case would be that the exhaust-gas mass flowrate would equal the rate at which the propellant burned. In equation (10.4) we have used the symbol V_E to denote the velocity of the exhaust gas at exit from the nozzle and F for the net force exerted on the fluid within the control volume.

If the pressure of the exhaust gas at exit from the rocket engine is p_E and S is the force exerted on the fluid within the control volume by the interior surfaces of the rocket engine, including the combustion chamber and the exhaust nozzle, then

$$F = S - p_E A_E \tag{10.5}$$

where A_E is the cross-sectional area of the nozzle at exit.

The condition for static equilibrium is

$$T - S + BA_E = 0 \tag{10.6}$$

where B is the ambient pressure and, as shown in Figure 10.2(b), we have used the symbol T for the restraining force which would be required to keep the rocket engine in place on a test bed, and this would be precisely the thrust exerted on a moving rocket-powered vehicle such as the space shuttle.

We can now combine equations (10.4), (10.5) and (10.6) to yield the final result for the thrust

$$T = \dot{m}V_E + (p_E - B)A_E \qquad (10.7)$$

which shows that thrust arises due to both the momentum flux of the exhaust-gas and also the exhaust pressure. Since the exhaust-gas flow is highly compressible, we cannot use Bernoulli's equation to connect p_E and V_E: this is a topic of compressible flow theory which is outside the scope of this text. However, the continuity equation (6.1) remains valid so that

$$\dot{m} = \rho V_E A_E$$

and we can write equation (10.7) as

$$T = \rho_E V_E^2 A_E + (p_E - B)A_E$$

where ρ_E is the exhaust-gas density at exit from the exhaust-gas nozzle. It is usually permissible to treat the exhaust gas as an ideal gas (see section 2.5), so that ρ_E can be calculated from the gas pressure p_E and absolute temperature (i.e. in degrees kelvin) T_E according to

$$p_E = \rho_E R T_E \qquad (2.6)$$

where R is the specific gas constant which depends upon the molecular weight of the gas but can often be taken as having the value $287 \, m^2/s^2.K$, which is appropriate for air.

EXAMPLE E10.2
The gas leaving the exhaust nozzle of a rocket engine has a temperature of 1500°C, a pressure of 1.7 bar and a velocity of 1800 m/s. If the exit diameter of the exhaust nozzle is 2.3 m and the ambient pressure is 0.1 bar, calculate the thrust developed by the engine. The exhaust gas can be assumed to behave as an ideal gas with a specific gas constant of $250 \, m^2/s^2.K$.

SOLUTION
$T_E = 1500 + 273 = 1773 \, K; \quad p_E = 1.7 \times 10^5 \, Pa; \quad R = 250 \, m^2/s^2.K;$
$V_E = 1800 \, m/s; \quad D_E = 2.3 \, m; \quad B = 10^4 \, Pa$

We first calculate the exhaust-gas density using the ideal gas law

$$\rho_E = \frac{p_E}{RT_E} = \frac{1.7 \times 10^5}{250 \times 1773} = 0.384 \, kg/m^3$$

and the exit cross-sectional area from

$$A_E = \tfrac{1}{4}\pi D_E^2 = \frac{\pi \times 2.3^2}{4} = 4.15 \text{ m}^2$$

The exhaust-gas momentum flux is then

$$\dot{M}_E = \rho_E V_E^2 A_E = 0.384 \times 1800^2 \times 4.15 = 5.16 \times 10^6 \text{ N} \quad \text{or} \quad 5.16 \text{ MN}$$

the pressure force is

$$(p_E - B)A_E = (1.7 \times 10^5 - 10^4) \times 4.15 = 6.6 \times 10^5 \text{ N} \quad \text{or} \quad 0.66 \text{ MN}$$

and the total thrust is

$$T = 5.16 + 0.66 = 5.82 \text{ MN}.$$

Comment: Both the momentum flux and the pressure force contribute significantly to the overall thrust. Since the ambient pressure decreases with altitude, the pressure force contribution will increase correspondingly and provide an increasing proportion of the overall thrust. The values used in this example are typical for a large liquid-propellant rocket engine such as one of the three main engines which power the space shuttle. The combined thrust provided by the two solid-propellant boosters is about five times that of the main engines.

10.3 Turbojet-engine thrust

A simplified cross-section of a basic jet engine, also referred to as a turbojet or a gas turbine, is shown in Figure 10.3(a). The function of a jet engine for propulsion purposes is to take in air from the surroundings and increase its momentum in a three-stage process: compression, combustion, expansion. Kerosene, a liquid fuel, is injected into combustion chambers where it burns in the air which has been raised to high pressure by an axial compressor. The power required to drive the compressor is produced by the expansion of hot gas flowing from the combustion chambers through the axial-flow turbine.

As for the rocket engine, the fluid control volume shown in Figure 10.3(b) corresponds to the interior of the engine. In the following equations we have used the subscript I to denote airflow conditions at the engine intake, E the exhaust-gas conditions at exit and F the fuel. As always in a steady-flow problem, it is convenient to consider flow relative to the engine, i.e. the engine is regarded as being stationary with the air flowing towards and past it at the speed of the aircraft V_A.

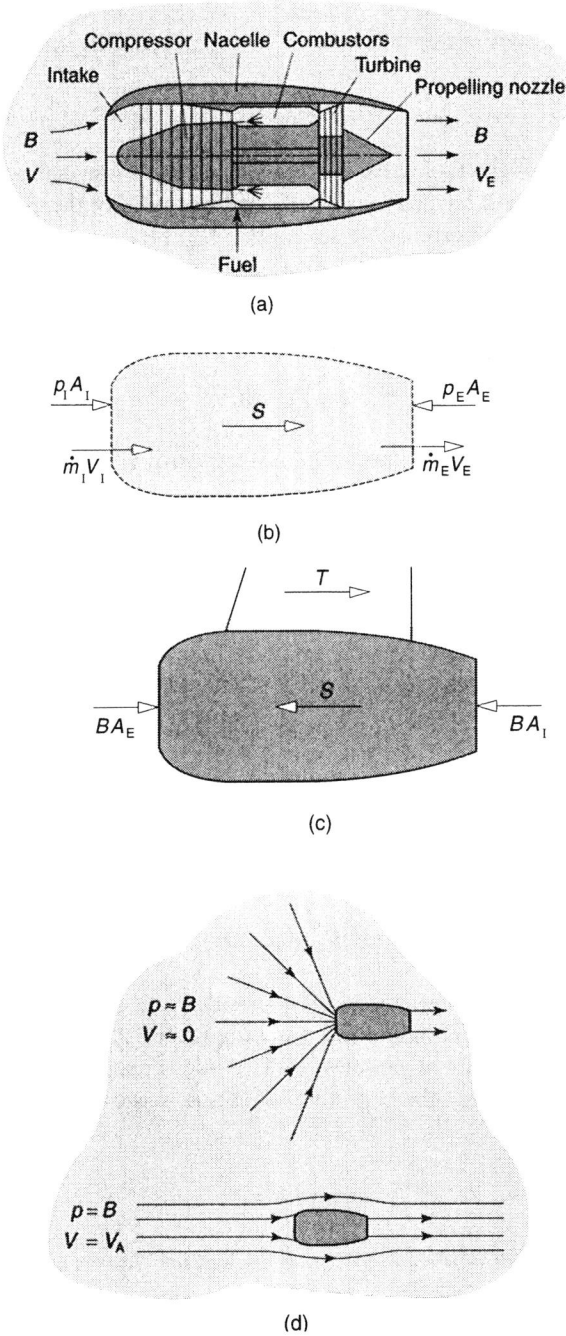

Figure 10.3 *Turbojet engine: (a) configuration; (b) fluid control volume; (c) forces acting on engine; (d) flow into and out of the engine*

Overall mass conservation requires that the mass flowrate of the exhaust gas $\dot{m}_E$ must equal the sum of the flowrates of air $\dot{m}_I$ and fuel $\dot{m}_F$ into the engine, i.e.

$$\dot{m}_E = \dot{m}_I + \dot{m}_F. \tag{10.8}$$

Since the velocity of the fuel at inlet to the engine V_F is low compared with that of the air V_I (which will be close to that of cruising speed of the aircraft), as with the rocket engine it is permissible to neglect the momentum flowrate of the inflowing fuel, $\dot{m}_F V_F$, compared with that of the air, $\dot{m}_I V_I$.

The momentum equation applied to the fluid flowing through the control volume can be written as

$$F = \dot{m}_E V_E - \dot{m}_I V_I \tag{10.9}$$

where the net force F acting on the fluid in the control volume is given by

$$F = S - p_E A_E + p_I A_I. \tag{10.10}$$

In this case the force S represents the net result of all the complex processes taking place within the engine.

The thrust T is again equal to the reaction force applied by the airframe to which the engine is connected, so that for static equilibrium of the engine (see Figure 10.3(c))

$$T - S - BA_I + BA_E = 0. \tag{10.11}$$

We can now combine equations (10.9) to (10.11) to give the fundamental thrust equation for jet propulsion

$$T = (\dot{m}_I + \dot{m}_F)V_E - \dot{m}_I V_I + (p_E - B)A_E - (p_I - B)A_I \tag{10.12}$$

and we now need to assign values to the pressures p_I and p_E. For subsonic conditions, it is reasonable to assume that the pressure p_E at the propelling nozzle outlet is equal to the ambient pressure B. The inlet pressure p_I is less straightforward to deal with. Since the continuity equation is again valid, for the intake airflow we have

$$\dot{m}_I = \rho_I A_I V_I$$

from which we can evaluate V_I if we know the air flowrate $\dot{m}_I$ and the air density ρ_I. The airflow approaching the intake may be assumed to satisfy Bernoulli's equation, so that p_I can be calculated from

$$p_0 = B + \tfrac{1}{2}\rho_I V_A^2 = p_I + \tfrac{1}{2}\rho_I V_I^2$$

where V_A represents the airspeed of the aircraft which must be the velocity of the airflow relative to the engine far upstream of the aircraft where the air pressure is equal to the ambient pressure B. Bernoulli's equation provides a more satisfactory way to specify the inlet pressure p_I than the commonly made assumption that $p_I = B$. As we can see, if the aircraft velocity V_A is close to that of the airflow entering the engine V_I, then the assumption that $p_I = B$ is justified. However, in the case of an engine on a test bed or for an aircraft just before takeoff, the air far from the engine will be at rest so that $V_A = 0$ and it is clear that air must be sucked into the engine by reducing p_I to a value much less than B. The two situations are illustrated in Figure 10.3(d). In other circumstances, the airspeed of the aircraft may exceed V_I, and if this occurs then p_I is greater than B and the engine benefits from a ram effect.

If we combine Bernoulli's equation with the continuity equation, then

$$p_I - B = \tfrac{1}{2} \rho_I \left(V_A^2 - \frac{\dot{m}_I^2}{\rho_I^2 A_I^2} \right). \tag{10.13}$$

To complete the analysis, we introduce the continuity equation for the exhaust flow

$$\dot{m}_E = \rho_E A_E V_E$$

which we can substitute into the thrust equation (10.12), together with equation (10.13) and the exhaust-pressure condition $p_E = B$, to give

$$T = \frac{(\dot{m}_I + \dot{m}_F)^2}{\rho_E A_E} - \left[1 + \left(\frac{V_A}{V_I} \right)^2 \right] \frac{\dot{m}_I^2}{2\rho_I A_I}. \tag{10.14}$$

The second term is called the *momentum drag* because it reduces the thrust below the value which would be obtained from the exhaust gas flow alone. If the airspeed V_A is equal to the airflow velocity into the engine V_I, we see from Bernoulli's equation that $p_I = B$ (i.e. the usual assumption made for p_I) and the thrust equation reduces to

$$T = \frac{(\dot{m}_I + \dot{m}_F)^2}{\rho_E A_E} - \frac{\dot{m}_I^2}{\rho_I A_I}.$$

For an engine at rest, however, $V_A = 0$ and we have

$$T = \frac{(\dot{m}_I + \dot{m}_F)^2}{\rho_E A_E} - \frac{\dot{m}_I^2}{2\rho_I A_I}.$$

the difference in the momentum drag arising because in this situation

$$p_1 = B - \tfrac{1}{2} \rho_1 V_1^2.$$

Although the turbojet engine is the simplest form of jet engine, it is widely used with thrust augmented by afterburners to power combat aircraft such as the Lockheed F-104S Starfighter which has a General Electric J79-GE-19 engine developing 79.6 kN thrust. More common than turbojet engines are now turbofan engines in which practically all of the thrust is generated by diverting most of the airflow through a large-diameter cowled fan. As we show in section 10.6, the most advanced turbofan engines, such as the Rolls-Royce Trent 890, can produce thrust in excess of 400 kN.

EXAMPLE 10.3

A turbojet engine has an inlet cross-sectional area of 2.9 m² and an exhaust nozzle with a cross-sectional area of 2.6 m². The engine powers an aircraft flying at a speed of 250 m/s at an altitude of 10 000 m where the ambient pressure is 0.265 bar and the ambient density is 0.41 kg/m³. The engine consumes fuel at a rate of 13.5 kg/s and operates with a fuel : air ratio of 0.05 : 1. The exhaust-gas density is 0.19 kg/m³. Calculate the air velocity and pressure at the engine intake, the velocity of the exhaust gas leaving the engine, and the thrust developed by the engine.

SOLUTION

$A_I = 2.9 \text{ m}^2$; $A_E = 2.6 \text{ m}^2$; $V_A = 250 \text{ m/s}$; $\rho_I = 0.41 \text{ kg/m}^3$;
$B = 2.65 \times 10^4 \text{ Pa}$; $\dot{m}_F = 13.5 \text{ kg/s}$; $\dot{m}_F/\dot{m}_I = 0.05$; $\rho_E = 0.19 \text{ kg/m}^3$

From the value for $\dot{m}_F$ and the fuel : air ratio we have

$$\dot{m}_I = \frac{13.5}{0.05} = 270 \text{ kg/s}.$$

We can now calculate the airflow velocity at the engine intake from the continuity equation, i.e.

$$V_I = \frac{\dot{m}_I}{\rho_I A_I} = \frac{270}{0.41 \times 2.9} = 227.1 \text{ m/s}.$$

In this case we see that V_I is slightly lower than V_A, so that from Bernoulli's equation the inlet pressure p_I must be slightly higher than the ambient pressure B:

$$p_I = B + \tfrac{1}{2}\,\rho_I(V_A^2 - V_I^2)$$
$$= 2.65 \times 10^4 + 0.5 \times 0.41 \times (250^2 - 227.1^2)$$
$$= 2.87 \times 10^4 \text{ Pa} \quad \text{or} \quad 0.287 \text{ bar.}$$

The values of the two terms in equation (10.14) are as follows

$$\frac{(\dot{m}_I + \dot{m}_F)^2}{\rho_E A_E} = 1.627 \times 10^5 \text{ N} \quad \text{or} \quad 162.7 \text{ kN}$$

$$\left[1 + \left(\frac{V_A}{V_I}\right)^2\right]\frac{\dot{m}_I^2}{2\rho_I A_I} = 6.78 \times 10^4 \text{ N} \quad \text{or} \quad 67.8 \text{ kN}$$

and the overall thrust is 94.9 kN.

The gas velocity at outlet we obtain from the continuity equation as

$$V_E = \frac{\dot{m}_I + m_F}{\rho_E A_E} = 574 \text{ m/s.}$$

10.4 Flow through a sudden enlargement

From the continuity equation we know that if there is an increase in the cross-sectional area of a duct through which fluid is flowing then, if the fluid density is constant, the fluid velocity will decrease. If the area increase is gradual, as for the diffuser of a Venturi tube, the assumption of frictionless flow is usually justified and Bernoulli's equation shows that the fluid static pressure will rise while the total pressure remains constant. However, in many practical applications the area increase has to take place suddenly; for example, if there is inadequate space for a well-designed diffuser or other design considerations, a sudden enlargement is advantageous. We now show how the static pressure recovery and stagnation pressure loss can be calculated for flow through a duct with a sudden area increase.

The flow geometry under consideration is shown in Figure 10.4(a) and the control volume in Figure 10.4(b). The duct downstream of the enlargement is taken to be cylindrical.* A key assumption in the analysis is that the static pressure is uniform across section ① where the flow enters the control volume and equal to the static pressure in the pipe just upstream of the enlargement. This

---

*The term *cylindrical* does not necessarily mean that the cross-section is circular, only that the duct is straight and has the same cross-section everywhere.

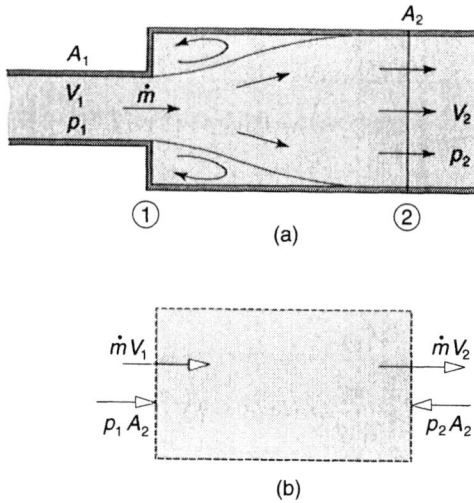

Figure 10.4 *Sudden enlargement: (a) flow geometry; (b) fluid control volume*

assumption, which in practice is well justified, implies that fluid enters the control volume as a jet with negligible streamline curvature. In the region immediately downstream of section ①, the jet is strongly affected by viscosity, becomes turbulent, diverges, and the flow rapidly adjusts to the downstream cross-section where the pressure and velocity can again be assumed to be uniform. So far as the overall momentum flux is concerned, the shear stress exerted on the flow by the duct wall is found to be negligible.

If we now apply the momentum equation to the flow through the control volume, we have

$$F = \dot{m}V_2 - \dot{m}V_1 = \dot{m}(V_2 - V_1).$$

Since the shear stress is negligible, and the duct downstream of the enlargement is parallel, the fluid–structure interaction force in this case must be zero and the force F acting on the fluid is due entirely to the pressures p_1 and p_2, i.e.

$$F = p_1 A_2 - p_2 A_2 = (p_1 - p_2)A_2.$$

We can now combine the two equations for F to find an equation for the pressure rise $p_2 - p_1$, i.e.

$$p_2 - p_1 = \frac{\dot{m}}{A_2}(V_1 - V_2). \qquad (10.15)$$

We introduce the continuity equation as

$$\dot{m} = \rho A_1 V_1 = \rho A_2 V_2$$

which allows equation (10.15) to be written as

$$p_2 - p_1 = \frac{\dot{m}^2}{\rho A_2}\left(\frac{1}{A_1} - \frac{1}{A_2}\right). \qquad (10.16)$$

Since the stagnation pressures upstream and downstream of the enlargement are given by

$$p_{0_1} = p_1 + \tfrac{1}{2}\rho V_1^2 \quad \text{and} \quad p_{0_2} = p_2 + \tfrac{1}{2}\rho V_2^2$$

the loss of stagnation pressure is given by

$$p_{0_1} - p_{0_2} = \tfrac{1}{2}\rho(V_1^2 - V_2^2) - (p_2 - p_1). \qquad (10.17)$$

If we again make use of the continuity equation to eliminate V_1 and V_2, we find

$$p_{0_1} - p_{0_2} = \frac{\dot{m}^2}{2\rho}\left(\frac{1}{A_1} - \frac{1}{A_2}\right)^2.$$

As we showed in section 7.5.2, pressure can be thought of as mechanical energy per unit volume, so this loss of stagnation pressure represents mechanical energy which is dissipated by viscous effects resulting in an increase in entropy and a small increase in the fluid temperature. Equation (10.17) can also be written as

$$p_{0_1} - p_{0_2} = \tfrac{1}{2}\rho V_1^2\left(1 - \frac{A_1}{A_2}\right)^2$$

i.e. the loss of stagnation pressure can also be viewed as a loss of kinetic energy. For the limiting case of a jet discharging to surroundings of effectively infinite extent ($A_2/A_1 \gg 1$) we see that the loss of stagnation pressure is equal to $\tfrac{1}{2}\rho V_1^2$, the dynamic pressure of the jet.

For a frictionless flow there is no loss of stagnation pressure and, from Bernoulli's equation, the pressure recovery $p_2 - p_1$ is given by

$$p_2 - p_1 = \tfrac{1}{2}\rho(V_1^2 - V_2^2) = \frac{\dot{m}^2}{2\rho A_2^2}\left[\left(\frac{A_2}{A_1}\right)^2 - 1\right]$$

which is larger than the pressure recovery for a sudden expansion (equation (10.16)) by the factor $\tfrac{1}{2}[(A_2/A_1) + 1]$. However, we emphasise that Bernoulli's equation does not apply to the sudden enlargement flow and the last result should be regarded as a basis for comparison.

EXAMPLE 10.4
Part of the exhaust system from an engine can be approximated by the axisymmetric contraction/enlargement shown in Figure E10.4. The upstream pipe and the outlet pipe both have a cross-sectional area of 3×10^{-3} m^2 and the contraction outlet has an area of 1.5×10^{-3} m^2. Calculate the differences in static pressure between sections ①, ② and ③, and the overall loss of stagnation pressure, for a gas of density 0.8 kg/m^3 with a mass flowrate of 0.2 kg/s. The entire flow may be regarded as incompressible and one dimensional, and the flow in the contraction as frictionless.

SOLUTION
$A_1 = 3 \times 10^{-3}$ m^2; $A_2 = 1.5 \times 10^{-3}$ m^2; $A_3 = 3 \times 10^{-3}$ m^2;
$\rho = 0.8$ kg/m^3; $\dot{m} = 0.2$ kg/s

We start by calculating the gas velocities at sections ①, ② and ③ from the continuity equation:

$$V_1 = \frac{\dot{m}}{\rho A_1} = \frac{0.2}{0.8 \times 3 \times 10^{-3}} = 83.3 \; m/s$$

$$A_2 = \tfrac{1}{2} A_1 \quad \text{so} \quad V_2 = 2V_1 = 186.7 \; m/s$$

$$A_3 = A_1 \quad \text{so} \quad V_3 = V_1 = 83.3 \; m/s.$$

Since the flow in the contraction is frictionless, Bernoulli's equation applies

$$p_1 + \tfrac{1}{2}\rho V_1^2 = p_2 + \tfrac{1}{2}\rho V_2^2$$

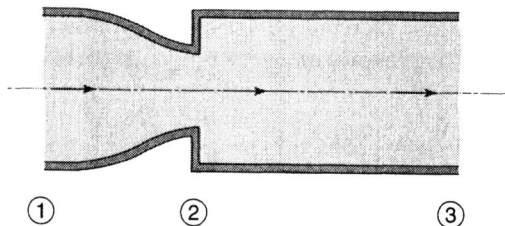

Figure E10.4

so that

$$p_1 - p_2 = \tfrac{1}{2}\rho(V_2^2 - V_1^2)$$
$$= 0.5 \times 0.8 \times (166.7^2 - 83.3^2)$$
$$= 8.33 \times 10^3 \text{ Pa} \quad \text{or} \quad 8.33 \text{ kPa}.$$

For the enlargement we can use the results of section 10.4, equation (10.15), so that

$$p_3 - p_2 = \frac{\dot{m}}{A_3}(V_2 - V_3)$$

$$= \frac{0.2(166.7 - 83.3)}{3 \times 10^{-3}}$$

$$= 5.56 \times 10^3 \text{ Pa} \quad \text{or} \quad 5.56 \text{ kPa}.$$

Since there is no change in stagnation pressure in the contraction, the entire stagnation pressure loss occurs between sections ② and ③. Here again we can use the results of section 10.4, equation (10.17), i.e.

$$p_{0_2} - p_{0_3} = \tfrac{1}{2}\frac{\dot{m}^2}{\rho}\left(\frac{1}{A_2} - \frac{1}{A_3}\right)^2$$

$$= \frac{0.5 \times 0.2^2}{0.8}\left(\frac{1}{1.5} - \frac{1}{3}\right)^2 \times 10^6$$

$$= 2.78 \times 10^3 \text{ Pa} \quad \text{or} \quad 2.78 \text{ kPa}.$$

10.5 Jet pump (or ejector or injector)

A basic jet pump consists of two concentric tubes as shown in Figures 1.13 and 10.5(a). A secondary flow occurs in the annulus surrounding the inner tube due to fluid being drawn into the high-speed jet of fluid discharging from the central pipe – a process known as entrainment. Vigorous mixing takes place between the primary and secondary streams to produce a homogeneous exit flow. In practice, for liquids, the primary jet flow usually discharges into the body of the pump through a convergent–nozzle designed to produce a high-speed, low-pressure jet flow. For gas flows a convergent-divergent nozzle to produce a low-pressure supersonic jet is more common. The design of the outer tube depends upon the application, but is often divergent to act as a diffuser if a high outlet pressure is

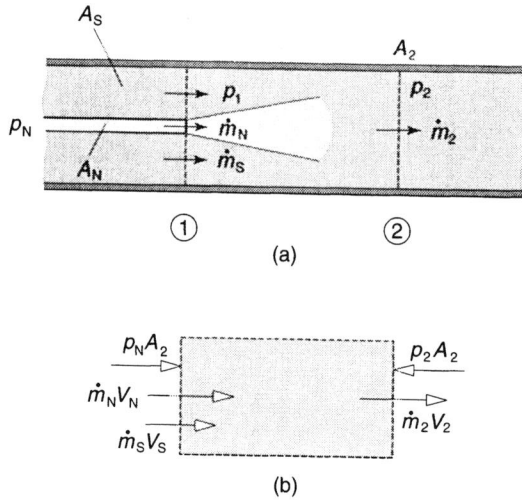

Figure 10.5 *Jet pump: (a) flow configuration; (b) fluid control volume*

required. A convergent exit nozzle (see section 10.5.2) to reduce the pressure and increase the momentum of the outflow is more suitable in thrust applications.

Flow machines such as the jet pump which have no moving parts are called fluidic devices. Jet pumps are cheap and robust, thereby requiring little maintenance, but are inefficient and can be noisy because of the high-speed of the primary flow. The low efficiency is often of little consequence because they are commonly used where the primary flow energy would otherwise be wasted. The numerous applications of jet pumps (also called ejectors and injectors) include water-aeration systems, water-jet aspirators (as commonly encountered in chemistry laboratories), feedwater pumps for boilers, air/gas mixers in domestic and industrial burners (the bunsen burner is an example), thrust augmentors, and pumps for sand, gravel, foodstuffs and other slurries.

For simplicity we shall consider an arrangement in which the main body of the pump is cylindrical with no change in cross-sectional area. If we apply the momentum equation to the fluid control volume shown in Figure 10.5(b) we have

$$(p_N - p_2)A_2 = (\dot{m}_N + \dot{m}_S)V_2 - \dot{m}_N V_N - \dot{m}_S V_S \qquad (10.18)$$

where the subscript S denotes the secondary flow, N the conditions at the exit of the injector tube and 2 refers to section ②. It should be noted that an important difference between the analysis of the jet pump and that for the sudden expansion, the rocket engine and the turbojet engine is that now we have two streams entering the control volume. The pressure at section ① is assumed to be the same in both streams and equal to the pressure p_N of the jet flow at exit from the injector tube. Since there may be a large velocity difference between the two streams (i.e. $V_N \gg V_S$), equation (10.18) accounts separately for the momentum flowrate of each stream. As we did for the sudden enlargement, in writing equation (10.18), we have neglected any shear stresses acting on the fluid at the outer tube wall, so that the only force exerted on the fluid in the control volume is due the pressure difference $(p_N - p_2)$.

Since the primary and secondary streams mix and leave the control volume as a single stream, overall mass-conservation requires that

$$\dot{m}_2 = \dot{m}_N + \dot{m}_S$$

and, for a cylindrical geometry, we have the area relationship

$$A_2 = A_N + A_S.$$

The continuity equation can be applied separately to both primary and secondary streams at section ① and also to the flow at section ②, where the two streams are assumed to be completely mixed and the velocity uniform. For simplicity, both streams are assumed to have the same density ρ, so that

$$\dot{m}_N = \rho A_N V_N$$
$$\dot{m}_S = \rho A_S V_S$$

and

$$\dot{m}_2 = \dot{m}_N + \dot{m}_S = \rho A_2 V_2.$$

To go further we need to specify which quantities can be regarded as known and which to be calculated, and this will depend upon the application. For example, we may wish to calculate the mass flowrate of the secondary flow $\dot{m}_S$ for a pump of given dimensions (i.e. A_N and A_S specified), a known fluid (density ρ), and given pressures p_N and p_2. An alternative might be to calculate p_2 for a given overall flowrate, $\dot{m}_2$. In both cases, we have sufficient information to carry out the calculations, as we shall demonstrate shortly. In other circumstances, we might be given the pressure of the secondary flow at a location upstream of section ① and the outlet pressure downstream of section ②, where the area is different from A_2. In such situations we need to introduce further assumptions. For example, for a pump completely submerged in a liquid, it might be appropriate to assume that the secondary flow up to section ① is frictionless and use Bernoulli's equation to relate p_N, V_S and the ambient fluid pressure B. A similar approach could be used to relate p_2 and V_2 to the conditions at outlet from a nozzle or diffuser downstream of section ②. We shall also give an example of this situation. It should be apparent that many other variations on this problem are possible.

10.5.1 *Jet pump with specified mass flowrates*

We first make use of the continuity equation for each flow stream to replace the velocities in equation (10.18) by the corresponding flowrates and cross-sectional areas, so that we have

$$(p_N - p_2)A_2 = \frac{1}{\rho}\left[\frac{(\dot{m}_N + \dot{m}_S)^2}{A_2} - \frac{\dot{m}_N^2}{A_N} - \frac{\dot{m}_S^2}{A_S}\right].$$

This equation is easily rearranged to give the pressure rise $p_2 - p_N$ as

$$p_2 - p_N = \frac{1}{\rho A_2}\left[\frac{\dot{m}_N^2}{A_N} + \frac{\dot{m}_S^2}{A_2 - A_N} - \frac{(\dot{m}_N + \dot{m}_S)^2}{A_2}\right].$$

It should be apparent that if p_2 is known, then this equation can be used to calculate $\dot{m}_S$ if all other quantities are specified.

EXAMPLE E10.5
A jet pump with the configuration shown in Figure 10.5(a) has been designed to raise the pressure of a gas of density 1.1 kg/m^3. The inner tube has a cross-sectional area of 5×10^{-4}m^2 while that of the outer tube is 8×10^{-3} m^2. The primary flowrate is 0.15 kg/s and the secondary flowrate 0.5 kg/s. Calculate the pressure increase produced by the pump.

SOLUTION
$\rho = 1.1$ kg/m^3; $A_N = 5 \times 10^{-4}$ m^2; $A_2 = 8 \times 10^{-3}$ m^2;
$\dot{m}_N = 0.15$ kg/s; $\dot{m}_S = 0.5$ kg/s

From the equation for $p_2 - p_N$ we have

$$p_2 - p_N = \frac{1}{1.1 \times 8 \times 10^{-3}}\left(\frac{0.15^2}{5 \times 10^{-4}} + \frac{0.5^2}{7.5 \times 10^{-3}} - \frac{0.65^2}{8 \times 10^{-3}}\right)$$

$$= 2900 \text{ Pa} \quad \text{or} \quad 2.9 \text{ kPa.}$$

Comment: If this pressure increase were specified then the equation for $p_2 - p_N$ could be used to calculate $\dot{m}_S$ as

$$2900 \times 1.1 \times 8 \times 10^{-3} = \frac{0.15^2}{5 \times 10^{-4}} + \frac{\dot{m}_S^2}{7.5 \times 10^{-3}} - \frac{(\dot{m}_S + 0.15)^2}{8 \times 10^{-3}}$$

which can be simplified by multiplying both sides by $7.5 \times 10^{-3} \times 8$ to give

$$1.531 = 2.7 + 8\dot{m}_S^2 - 7.5(\dot{m}_S + 0.15)^2$$

and, after further simplification, we have the quadratic equation for $\dot{m}_S$

$$\dot{m}_S^2 - 4.5\dot{m}_S + 2 = 0.$$

The two solutions are $\dot{m}_S = 0.5$ kg/s and 4.0 kg/s. The first solution is the one which corresponds with our original problem, while the second corresponds to a secondary flowspeed of 455 m/s and would have to be excluded since this represents a supersonic flow condition.

The situation becomes more difficult if we need to calculate A_2 given all other quantities because the equation for $p_2 - p_N$ results in a cubic equation for A_2 which is much more difficult to solve than a quadratic equation.

10.5.2 *Jet pump with specified external pressures*

We consider the configuration shown in Figure 10.6(a) in which a jet pump with a convergent exit nozzle is used to generate a high-speed flow of a liquid. We shall assume that the secondary flow is frictionless upstream of section ① and that the mixed flow is frictionless downstream of section ② so that we can use Bernoulli's equation as follows

$$p_0 = p_N + \tfrac{1}{2}\,\rho V_S^2$$

and

$$p_2 + \tfrac{1}{2}\,\rho V_2^2 = p_E + \tfrac{1}{2}\,\rho V_E^2.$$

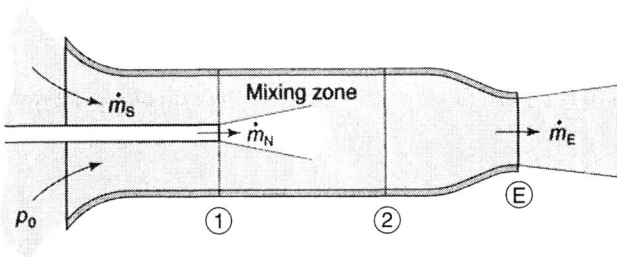

(a)

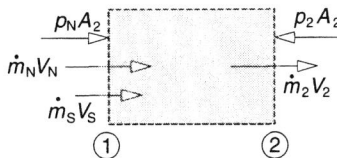

(b)

Figure 10.6 *Jet pump with convergent exit nozzle: (a) flow geometry; (b) fluid control volume*

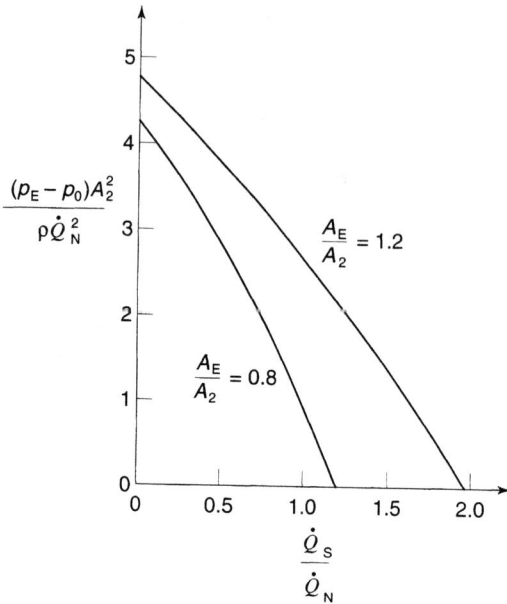

Figure 10.6 *(c) performance curves*

In the first of these two equations it has been assumed that the secondary flow is drawn from a reservoir with stagnation pressure p_0 and we have again made use of the pressure condition at section ①, $p_N = p_S$. The exit-flow condition for the second equation corresponds to a jet discharging into the surroundings at a pressure p_E.

The continuity equation is needed to relate the velocities at section ② and the exit, i.e.

$$\dot{Q}_E = A_2 V_2 = A_E V_E$$

where A_E is the cross-sectional area of the pump exit and $\dot{Q}_E$ is the total volumetric flowrate through the pump.

The momentum equation is the same as before, i.e.

$$(p_N - p_2)A_2 = (\dot{m}_N + \dot{m}_S)V_2 - \dot{m}_N V_N - \dot{m}_S V_S \qquad (10.18)$$
$$= \rho[(\dot{Q}_N + \dot{Q}_S)V_2 - \dot{Q}_N V_N - \dot{Q}_S V_S]$$

where we have now introduced the volumetric flowrates $\dot{Q}_N$ and $\dot{Q}_S$, i.e.

$$\dot{Q}_N = \frac{\dot{m}_N}{\rho} = A_N V_N \quad \text{and} \quad \dot{Q}_S = \frac{\dot{m}_S}{\rho} = A_S V_S.$$

There is little to choose between working with the mass flowrate and the volumetric flowrate for constant-density flows, although the latter is probably more usual for liquid flows.

If we use the continuity equation to eliminate the velocities from the momentum equation, we have

$$\frac{(p_2 - p_N)A_2}{\rho} = \frac{\dot{Q}_N^2}{A_N} + \frac{\dot{Q}_S^2}{A_S} - \frac{(\dot{Q}_N + \dot{Q}_S)^2}{A_2}.$$

The two equations for p_0 and p_E can then be used to replace $p_2 - p_N$ by

$$p_2 - p_N = p_E - p_0 + \tfrac{1}{2}\,\rho(V_E^2 - V_2^2 + V_S^2)$$

$$= p_E - p_0 + \tfrac{1}{2}\,\rho\left[\left(\frac{\dot{Q}_N + \dot{Q}_S}{A_E}\right)^2 - \left(\frac{\dot{Q}_N + \dot{Q}_S}{A_2}\right)^2 + \left(\frac{\dot{Q}_S}{A_S}\right)^2\right].$$

If we substitute this result for $p_2 - p_N$ in the momentum equation, after some algebra we find that the overall pressure rise is given by

$$\frac{(p_E - p_0)A_2^2}{\rho \dot{Q}_N^2} = \frac{A_2}{A_N} - \frac{1}{2} - \frac{1}{2}\left(\frac{A_2}{A_E}\right)^2 - \frac{\dot{Q}_S}{\dot{Q}_N}\left[1 + \left(\frac{A_2}{A_E}\right)^2\right]$$

$$+ \left(\frac{\dot{Q}_S}{\dot{Q}_N}\right)^2\left[\frac{A_2}{A_S} - \frac{1}{2} - \frac{1}{2}\left(\frac{A_2}{A_S}\right)^2 - \frac{1}{2}\left(\frac{A_2}{A_E}\right)^2\right].$$

It can be seen that the final result here is in non-dimensional form and all quantities on the right-hand side appear as ratios. As we argued in Chapter 3, it is always more convenient to present any result, whether theoretical or experimental, in non-dimensional form. In the present case the advantage is that the geometry of a range of geometrically similar pumps can be represented by the two parameters A_N/A_2 and A_E/A_2 (since $A_S = A_2 - A_N$, $A_S/A_2 = 1 - A_N/A_2$), and the non-dimensional pressure rise calculated for a range of values for the ratio $\dot{Q}_S/\dot{Q}_N$ to produce a single performance curve. Figure 10.6(c) shows two such curves, both for $A_N/A_2 = 0.18$ but for two different values of A_E/A_2, 0.8 corresponding to a convergent nozzle and 1.2 representing a diffuser. Although we carried out the analysis with the nozzle arrangement in mind, it applies equally well to the diffuser situation. As we should expect, for any given value of $\dot{Q}_S/\dot{Q}_N$, the results show that the diffuser produces a higher exit pressure than the nozzle. We also see that the jet pump becomes increasingly effective as $\dot{Q}_S/\dot{Q}_N$ is reduced, i.e. as the primary flowrate is increased (which would also require an increase in the stagnation pressure p_0).

EXAMPLE E10.6
A jet pump has dimensions corresponding to the following cross-sectional areas: $A_N = 5 \times 10^{-4}\,\text{m}^2$, $A_2 = 8 \times 10^{-3}\,\text{m}^2$ and $A_E = 6 \times 10^{-3}\,\text{m}^2$. The pump operates completely submerged in water at a depth of 5 m where the water pressure is 1.51 bar, which can be taken as both the stagnation pressure of the secondary flow and also the pump outlet pressure. If the primary mass flowrate is 25 kg/s, calculate the secondary and overall mass flowrates and also the pressures p_N and p_2.

SOLUTION
$A_N = 5 \times 10^{-4}\,\text{m}^2$; $A_2 = 8 \times 10^{-3}\,\text{m}^2$; $A_E = 6 \times 10^{-3}\,\text{m}^2$;
$p_0 = p_E = 1.51 \times 10^5\,\text{Pa}$; $\dot{m}_S = 25\,\text{kg/s}$; $\rho = 1000\,\text{kg/m}^3$

From the areas we can find the area ratios as

$$\frac{A_N}{A_2} = 0.0625;\quad \frac{A_S}{A_2} = \frac{A_2 - A_N}{A_2} = 0.9375;\quad \frac{A_E}{A_2} = 0.75$$

so that the area terms in the equation for $p_E - p_0$ are

$$\frac{A_2}{A_N} - \tfrac{1}{2} - \tfrac{1}{2}\left(\frac{A_2}{A_E}\right)^2 = 14.61$$

$$1 + \left(\frac{A_2}{A_E}\right)^2 = 2.78$$

and

$$\frac{A_2}{A_S} - \tfrac{1}{2} - \tfrac{1}{2}\left(\frac{A_2}{A_S}\right)^2 - \tfrac{1}{2}\left(\frac{A_2}{A_E}\right)^2 = -0.89.$$

Since $p_E = p_0$ in this case, we have

$$0.89\left(\frac{\dot{Q}_S}{\dot{Q}_N}\right)^2 + 2.78\,\frac{\dot{Q}_S}{\dot{Q}_N} - 14.61 = 0,$$

i.e. a quadratic equation which we can solve for $\dot{Q}_S/\dot{Q}_N$ to give

$$\frac{\dot{Q}_S}{\dot{Q}_N} = 2.78.$$

The second root is negative and so ruled out on physical grounds.
Since the flow is incompressible, the mass flowrates are in the same ratio

as the volumetric flowrates so that

$$\dot{m}_S = 2.78 \times 25 = 69.5 \text{ kg/s}$$

and the total mass flowrate is

$$\dot{m}_E = \dot{m}_N + \dot{m}_S = 94.5 \text{ kg/s}.$$

In the derivation of the equation for $p_E - p_0$ it was assumed that the flow upstream of section ① is frictionless. We can therefore use Bernoulli's equation to calculate the pressure p_N from

$$p_N = p_0 - \tfrac{1}{2}\, \rho V_S^2.$$

From the continuity equation, the velocity of the secondary flow upstream of section ① is

$$V_S = \frac{\dot{m}_S}{\rho A_S} = \frac{69.5}{1000 \times 7.5 \times 10^{-3}} = 9.27 \text{ m/s}$$

and so

$$p_N = (1.51 \times 10^5) - (0.5 \times 1000 \times 9.27^2) = 1.08 \times 10^5 \text{ Pa} \quad \text{or} \quad 1.08 \text{ bar}.$$

To find p_2 we can again use Bernoulli's equation

$$p_2 + \tfrac{1}{2}\, \rho V_2^2 = p_E + \tfrac{1}{2}\, \rho V_E^2$$

so that

$$p_2 = p_E + \tfrac{1}{2}\, \rho(V_E^2 - V_2^2).$$

From the continuity equation

$$V_E = \frac{\dot{m}_E}{\rho A_E} = \frac{94.5}{1000 \times 6 \times 10^{-3}} = 15.75 \text{ m/s}$$

and

$$V_2 = \frac{\dot{m}_E}{\rho A_2} = \frac{94.5}{1000 \times 8 \times 10^{-3}} = 11.81 \text{ m/s}.$$

Hence

$$p_2 = (1.51 \times 10^5) + (0.5 \times 1000)(15.75^2 - 11.81^2)$$
$$= 2.05 \times 10^5 \text{ Pa} \quad \text{or} \quad 2.05 \text{ bar}.$$

10.6 **Turbofan-engine thrust**

We return at this point to consider the thrust produced by a jet engine, in this case a turbofan engine such as that shown schematically in Figures 1.8 and 10.7(a). Most of the air flowing through a turbofan engine bypasses the combustors and turbine stages: bypass ratios of 4 : 1 to 6 : 1 are typical of large modern turbofan

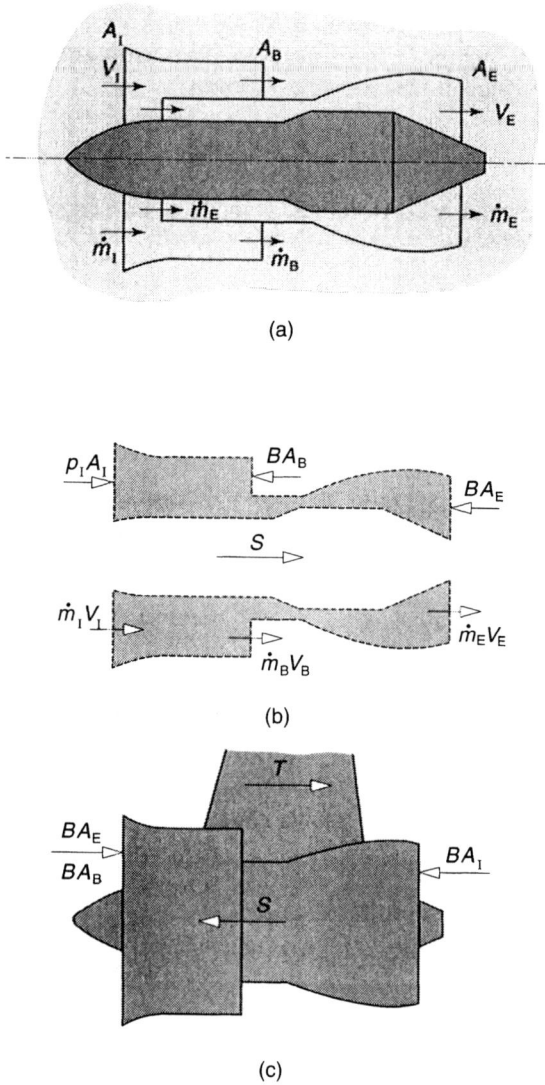

Figure 10.7 *Turbofan engine: (a) flow configuration; (b) fluid control volume; (c) forces acting on engine*

engines such as the Rolls-Royce RB211 and Trent engines which power such aircraft as the Boeing 747 and Airbus A330. The overall thrust generated is a combination of the thrust produced by the ducted fan mounted ahead of the compressor stages and the hot exhaust jet. As with the jet pump, we now have two fluid streams to consider.

We consider the fluid control volume shown in Figure 10.7(b). The total mass flowrate into the engine is $\dot{m}_I$, the bypass flowrate is $\dot{m}_B$ and the fuel mass flowrate is $\dot{m}_F$. For optimum combustion the air : fuel ratio for a large modern turbofan engine is normally in the range 3×10^{-3} to 4×10^{-3}. Overall mass conservation requires that the exhaust mass flowrate m_E is given by

$$\dot{m}_E + \dot{m}_B = \dot{m}_I + \dot{m}_F.$$

If the air density at inlet to the engine is ρ_I, the air density at outlet from the bypass ducting is ρ_B, and the exhaust gas density is ρ_E, then from the continuity equation we have

$$\dot{m}_I = \rho_I A_I V_I, \quad \dot{m}_B = \rho_B A_B V_B \quad \text{and} \quad \dot{m}_E = \rho_E A_E V_E$$

from which the fluid velocities at inlet, bypass outlet and exhaust can be calculated given the cross-sectional areas A_I, A_B and A_E.

If F is the net force exerted on the fluid within the control volume (i.e. including both streams), then from the linear momentum equation we have

$$F = \dot{m}_B V_B + \dot{m}_E V_E - \dot{m}_I V_I.$$

If the air pressure at the front face of the fan is p_1 and both streams are assumed to leave the engine at ambient pressure B, then

$$F = S + p_I A_I - B(A_B + A_E)$$

where S is the fluid–structure interaction force, again including the effects of both fluid streams.

From Figure 10.7(c) we can see that the reaction force R required to satisfy the condition of static equilibrium for the engine is given by

$$R - S - BA_I + B(A_B + A_E) = 0$$

so that the thrust T, which must be equal in magnitude but opposite in direction to R, is obtained from

$$T = S + BA_I - B(A_B + A_E)$$
$$= \dot{m}_B V_B + \dot{m}_E V_E - \dot{m}_I V_I - (p_I - B)A_I.$$

The following numerical example, which is based upon figures for the Rolls-Royce Trent 890 turbofan engine, leads to a calculated thrust remarkably close to the value quoted by Rolls-Royce for this engine.

EXAMPLE E10.7

Air at a pressure of 0.8 bar with a density of 1.01 kg/m³ enters a turbofan engine at a rate of 1200 kg/s. The bypass ratio is 5.9 : 1 and the fuel flowrate is 4 kg/s. The bypass air leaves the engine with a density of 1.37 kg/m³ and the exhaust-gas density is 0.15 kg/m³. The ambient pressure is 1.01 bar. The cross-sectional areas are 6.13 m² at inlet, 0.89 m² for the exhaust nozzle and 2.65 m² for the bypass ducting. Calculate the thrust developed by the engine.

SOLUTION

$\dot{m}_I = 1200$ kg/s; $r = 5.9$; $\dot{m}_F = 4$ kg/s; $p_I = 8 \times 10^4$ Pa;
$\rho_I = 1.01$ kg/m³; $\rho_B = 1.37$ kg/m³; $\rho_E = 0.15$ kg/m³;
$p_B = p_E = 1.01 \times 10^5$ Pa; $A_I = 6.13$ m²; $A_B = 2.65$ m²; $A_E = 0.89$ m²

Since the bypass ratio $r = 5.9$, and if the bypass flowrate is $\dot{m}_B$, then the total airflow rate entering the engine must be given by

$$\dot{m}_I = \dot{m}_B + \frac{\dot{m}_B}{r} = 1200 \text{ kg/s}$$

from which

$$\dot{m}_B = 1026 \text{ kg/m}^3$$

and the exhaust flowrate $\dot{m}_E$ must be given by

$$\dot{m}_E = (\dot{m}_I - \dot{m}_B) + \dot{m}_F$$
$$= 1200 - 1026 + 4 = 178 \text{ kg/s.}$$

We can now use the continuity equation to find the gas velocities as follows:

$$V_I = \frac{\dot{m}_I}{\rho_I A_I} = \frac{1200}{1.01 \times 6.13} = 193.8 \text{ m/s}$$

$$V_B = \frac{\dot{m}_B}{\rho_B A_B} = \frac{1026}{1.37 \times 2.65} = 282.6 \text{ m/s}$$

and

$$V_E = \frac{\dot{m}_E}{\rho_E A_E} = \frac{178}{0.15 \times 0.89} = 1333 \text{ m/s.}$$

The thrust can now be calculated from

$$T = \dot{m}_B V_B + \dot{m}_E V_E - \dot{m}_I V_I - (p_I - B)A_I$$
$$= (1026 \times 282.6) + (178 \times 1333) - (1200 \times 193.8)$$
$$-[(8 \times 10^4) - (1.01 \times 10^5)] \times 6.13$$
$$= (2.9 \times 10^5) + (2.37 \times 10^5) - (2.33 \times 10^5) + (1.29 \times 10^5)\text{N}$$
$$= 4.23 \times 10^5 \text{ N} \quad \text{or} \quad 423 \text{ kN}.$$

10.7 Reaction force on a pipe bend

The six preceding sections of this chapter have dealt with flows where changes in velocity, pressure and momentum were a consequence of changes in cross-sectional area. We now consider situations where the flow geometry leads to a change in flow direction.

We start with a pipe bend which turns through an angle θ and at the same time changes in cross-sectional area. This is a more general version of the 90°-pipe bend discussed in the previous chapter. The flow configuration is shown in Figures 1.11 and 10.8(a), the fluid control volume in Figure 10.8(b) and the forces exerted on the bend in Figure 10.8(c). For convenience, the inflow at section ① is taken to be in the x-direction and the outflow at section ② to be at an angle θ to the x-direction.

The momentum equation applied to the fluid in the control volume gives

$$p_1 A_1 - p_2 A_2 \cos \theta - S_x = \dot{m}(V_2 \cos \theta - V_1)$$

for the x-direction, and

$$-p_2 A_2 \sin \theta + S_y = \dot{m} V_2 \sin \theta$$

for the y-direction. With experience, it is usually straightforward to decide whether the components of the fluid–structure interaction force S_x and S_y should be positive or negative. However, it is of no consequence if we choose incorrectly: the directions of both **S** and the reaction force **R** are results of the analysis. Also, because the inflow at section ① is in the x-direction, there is no contribution of either the pressure force $p_1 A_1$ or the momentum flux $\dot{m}V_1$ to the y-momentum equation. In contrast, unless the outflow at section ② is at 90° to the inflow, both $p_2 A_2$ and $\dot{m}V_2$ will contribute to the x-momentum equation.

From Figure 10.8(c) we can see that the condition of static equilibrium leads to

$$-R_x + S_x - BA_1 + BA_2 \cos \theta = 0$$

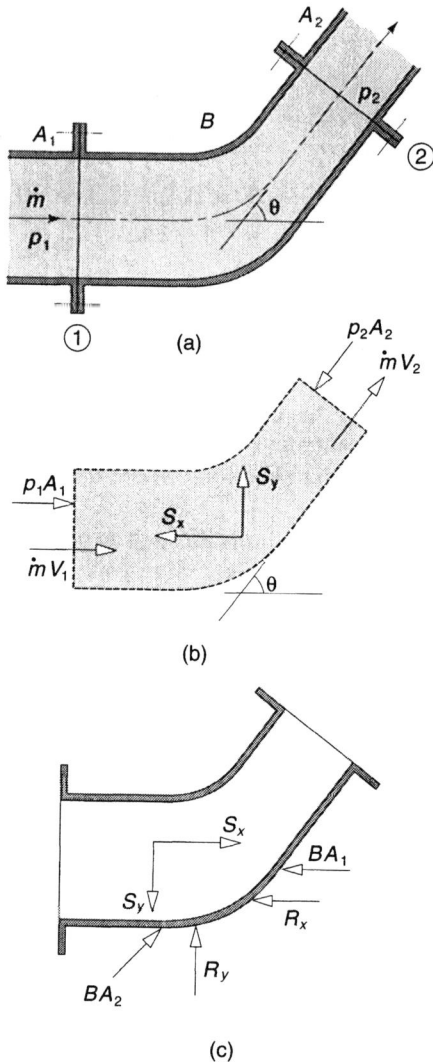

Figure 10.8 *Pipe bend: (a) flow configuration; (b) fluid control volume; (c) forces exerted*

and

$$R_y - S_y + BA_2 \sin \theta = 0.$$

We can now eliminate S_x to give an equation for R_x

$$R_x = \dot{m}(V_1 - V_2 \cos \theta) + (p_1 - B)A_1 - (p_2 - B)A_2 \cos \theta \qquad (10.19)$$

and eliminate S_y to give

$$R_y = \dot{m}V_2 \sin \theta + (p_2 - B)\sin \theta$$

for R_y. The magnitude of the net hydrodynamic reaction force $\mathbf{R}$ and its direction ξ are then found from

$$R = \sqrt{(R_x^2 + R_y^2)} \quad \text{and} \quad \xi = \tan^{-1}\left(\frac{R_y}{R_x}\right)$$

As we commented in section 9.4, the *reaction force* $\mathbf{R}$ depends upon the *gauge pressures* $p_1 - B$ and $p_2 - B$ rather than the absolute static pressures p_1 and p_2.

Up to this point the analysis has been quite general. To go further we need more information, for example about the pressures p_1 and p_2. A common outlet condition, though not the only one, is that the flow discharges at atmospheric pressure B so that $p_2 - B = 0$. If, in addition, the flow within the pipe bend can be assumed to be frictionless, then p_1 and p_2 (now equal to B) can be related using Bernoulli's equation, i.e.

$$p_1 + \tfrac{1}{2}\rho V_1^2 = B + \tfrac{1}{2}\rho V_2^2.$$

To combine Bernoulli's equation with the equation for R_x requires introduction of the continuity equation

$$\dot{m} = \rho A_1 V_1 = \rho A_2 V_2.$$

After some algebra we find

$$R_x = \frac{\dot{m}^2}{2\rho A_1}\left[\left(\frac{A_1}{A_2}\right)^2 - \frac{2A_1}{A_2}\cos\theta + 1\right]$$

$$= \tfrac{1}{2}\rho A_1 V_1^2\left[\left(\frac{A_1}{A_2}\right)^2 - \frac{2A_1}{A_2}\cos\theta + 1\right]$$

while the equation for R_y (with $p_2 - B = 0$) simplifies to

$$R_y = \frac{\dot{m}^2}{\rho A_2}\sin\theta$$

$$= \tfrac{1}{2}\rho A_1 V_1^2\frac{2A_1}{A_2}\sin\theta.$$

The second version of each of these equations for R_x and R_y has the form

$$\text{force} = \text{non-dimensional geometric factor} \times \text{dynamic pressure} \times \text{area}$$

much like the equation for drag force

$$\text{drag force} = \text{drag coefficient} \times \text{dynamic pressure} \times \text{area.}$$

As we noted at the end of section 10.1, this form is typical for many practical flow situations.

EXAMPLE E10.7
Liquid of density ρ flows through a pipe which turns through 180° and at the same time doubles in area. The static pressure before the bend is twice the ambient pressure B and the flow within the bend can be regarded as frictionless. Show that the external reaction force R required to restrain the bend is given by

$$R = \frac{9\dot{m}^2}{4\rho A} + 3BA$$

where $\dot{m}$ is the mass flowrate and A is the cross-sectional area ahead of the bend.

SOLUTION
We start by substituting $p_1 = 2B$, $A_1 = A$, $A_2 = 2A$, and $\theta = 180°$ in equation (10.19) for R_x (because $\theta = 180°$, R_y must be zero and so $R = R_x$), i.e.

$$R = \frac{3\dot{m}^2}{2\rho A} + BA + 2(p_2 - B)A$$

wherein we have also made use of the continuity equation which gives

$$V_1 = 2V_2 = \frac{\dot{m}}{\rho A}.$$

Then, from Bernoulli's equation, we have

$$p_2 = p_1 + \tfrac{1}{2}\rho V_1^2 - \tfrac{1}{2}\rho V_2^2$$

$$= 2B + \frac{\dot{m}^2}{2\rho A^2} - \frac{\dot{m}^2}{8\rho A^2}$$

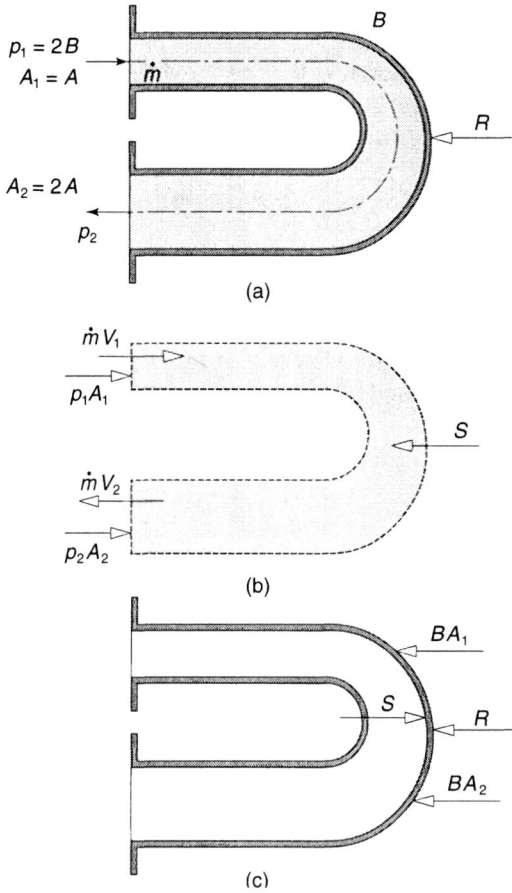

$p_1 = 2B$
$A_1 = A$
$\dot{m}$
B
R
$A_2 = 2A$
p_2
(a)

$\dot{m}\,V_1$
$p_1 A_1$
S
$\dot{m}\,V_2$
$p_2 A_2$
(b)

BA_1
S
R
BA_2
(c)

Figure E10.7 *(a) (b) (c)*

so that

$$p_2 - B = B + \frac{3\dot{m}^2}{8\rho A^2}.$$

If we substitute for $p_2 - B$ in the equation for R, we have

$$R = \frac{9\dot{m}^2}{4\rho A} + 3BA.$$

Although this is the required result, rather than starting by simply substituting in the equation for R_x (which would probably not be available in an examination), it would be better to carry out the analysis from first principles, as follows.

We first draw a diagram of the fluid control volume including the forces acting on the fluid and the momentum fluxes, as shown in Figure E10.7(b).

The momentum equation applied to the fluid control volume gives

$$-S + p_1 A_1 + p_2 A_2 = -\dot{m}V_2 - \dot{m}V_1.$$

Note that care has to be taken over the sign given to the term at outlet from the control volume to account properly for the fact that the duct has turned a full 180°.

From Figure E10.7(c), we see that the condition of static equilibrium applied to the pipe bend leads to

$$-R + S - BA_1 - BA_2 = 0$$

so that

$$R = \dot{m}(V_1 + V_2) + (p_1 - B)A_1 + (p_2 - B)A_2$$

which can be seen to lead to the same result as before.

10.8 Reaction force on a pipe junction

When we applied the momentum equation to the jet pump in section 10.5 we pointed out that because there were two streams entering the fluid control volume it was essential to account separately for the momentum flowrate of each stream. Since the pipe junction shown in Figure 10.9 has two outlets, each at a different angle to the flow direction at its inlet, we must now account separately not only for the mass and momentum flowrates of each outlet stream but also for the pressure force at each outlet.

The momentum equation applied to the control volume leads to

$$-S_x + p_1 A_1 - p_2 A_2 \cos \theta_2 - p_3 A_3 \cos \theta_3 = \dot{m}_2 V_2 \cos \theta_2 + \dot{m}_3 V_3 \cos \theta_3 - \dot{m}_1 V_1$$

and

$$S_y - p_2 A_2 \sin \theta_2 + p_3 A_3 \sin \theta_3 = \dot{m}_2 V_2 \sin \theta_2 - \dot{m}_3 V_3 \sin \theta_3.$$

The corresponding static-equilibrium conditions are

$$S_x - R_x - BA_1 + BA_2 \cos \theta_2 + BA_3 \cos \theta_3 = 0$$

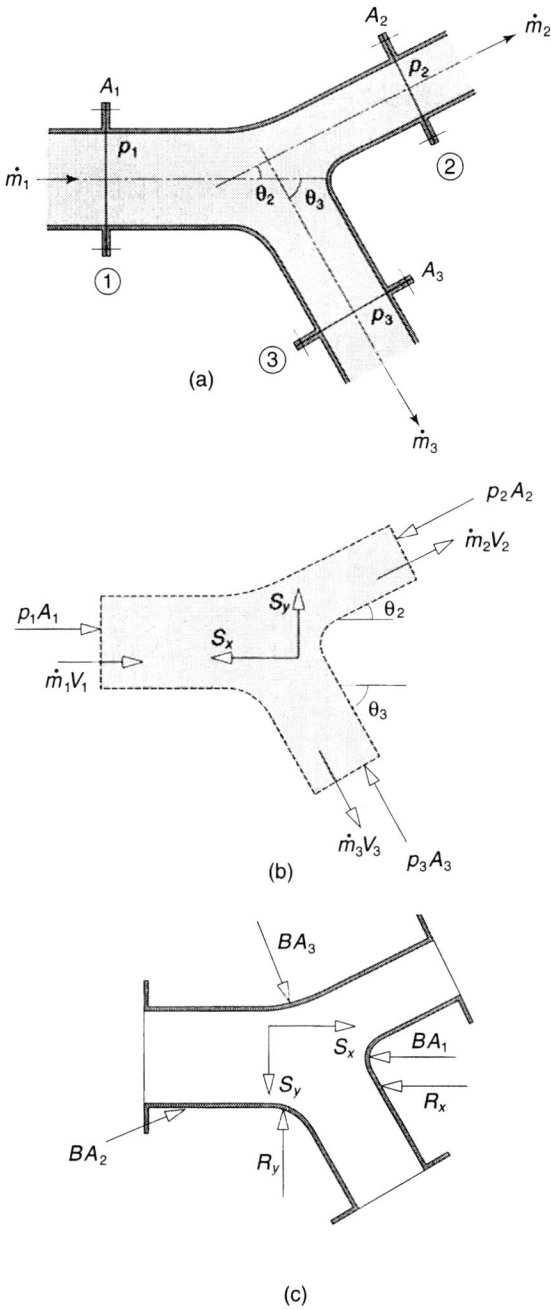

Figure 10.9 *Pipe junction: (a) flow geometry; (b) fluid control volume; (c) forces exerted*

and

$$R_y - S_y + BA_2 \sin \theta_2 - BA_3 \sin \theta_3 = 0.$$

The preceding equations can be combined as usual to eliminate S_x and S_y with the result

$$R_x = \dot{m}_1 V_1 - \dot{m}_2 V_2 \cos \theta_2 - \dot{m}_3 V_3 \cos \theta_3 + (p_1 - B)A_1 - (p_2 - B)A_2 \cos \theta_2$$
$$- (p_3 - B)A_3 \cos \theta_3$$

and

$$R_y = \dot{m}_2 V_2 \sin \theta_2 - \dot{m}_3 V_3 \sin \theta_3 + (p_2 - B)A_2 \sin \theta_2 - (p_3 - B)A_3 \sin \theta_3.$$

Overall mass conservation requires that

$$\dot{m}_1 = \dot{m}_2 + \dot{m}_3$$

and the continuity equation applied to each of the three streams gives

$$\dot{m}_1 = \rho \dot{Q}_1 = \rho A_1 V_1$$
$$\dot{m}_2 = \rho \dot{Q}_2 = \rho A_2 V_2$$

and

$$\dot{m}_3 = \rho \dot{Q}_3 = \rho A_3 V_3.$$

To proceed further we need information about the pressures p_1, p_2 and p_3. An interesting situation arises if the flow can be assumed to be frictionless since this implies that the stagnation pressure of all three streams must be the same. From Bernoulli's equation we then have

$$p_0 = p_1 + \tfrac{1}{2} \rho V_1^2$$
$$= p_2 + \tfrac{1}{2} \rho V_2^2$$
$$= p_3 + \tfrac{1}{2} \rho V_3^2.$$

It should be apparent that by properly accounting for all the relevant pressure forces and momentum flowrates, we could generalise the pipe junction analysis to include any number of inlets and outlets. It is at this point that any further development would be best dealt with by writing a computer program.

10.9 Flow through a cascade of guidevanes

A cascade (or set) of curved guidevanes (or turning vanes), such as is shown in Figures 1.14 and 10.10(a), is frequently used in wind tunnels, water channels,

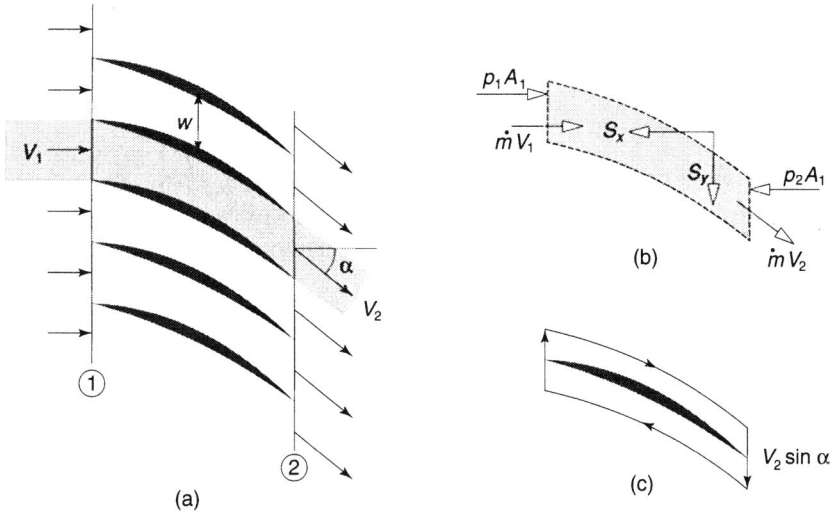

Figure 10.10 *Flow through a cascade of guidevanes: (a) flow geometry; (b) fluid control volume; (c) circulation loop*

air-conditioning ducts, etc., to change the direction of a gas or liquid stream. In these applications the guidevanes may well be formed from sheet metal. In more demanding applications, such as for the stator (i.e. non-rotating blading) stages of axial-flow turbines and compressors, the guidevane profile is more likely to resemble that of an aerofoil. Guidevanes are also used in a radial configuration, called a nozzle ring, to create a swirling flow into the rotors of hydraulic torque converters, centrifugal compressors and hydraulic (i.e. hydro-electric power) turbines. As we shall demonstrate, consideration of the flow through a cascade also gives some insight into aerofoil lift and drag.

We consider flow through a single channel between adjacent guidevanes. If the guidevane pitch (i.e. separation distance) is w and the span is s, the cross-sectional area at inlet to the channel (section ①) is given by

$$A_1 = ws.$$

If the deflection angle is α, the effective width of a channel at outlet (section ②), taken normal to the flow, must be $w \cos \alpha$ and the effective outlet area is

$$A_2 = w \cos \alpha \, s = A_1 \cos \alpha.$$

From the continuity equation

$$\dot{m} = \rho \dot{Q} = \rho A_1 V_1 = \rho A_2 V_2$$

we see that

$$V_2 = \frac{V_1}{\cos \alpha}.$$

Since $\cos \alpha < 1$, it is apparent that one effect of the guidevanes is to accelerate the fluid flowing through the cascade. If we assume the flow to be frictionless, we can use Bernoulli's equation to calculate the corresponding pressure drop, i.e.

$$p_0 = p_1 + \tfrac{1}{2} \rho V_1^2 = p_2 + \tfrac{1}{2} \rho V_2^2$$

from which

$$p_1 - p_2 = \tfrac{1}{2} \rho V_1^2 \tan^2 \alpha.$$

If we apply the momentum equation to the fluid control volume shown in Figure 10.9(b), we have

$$-S_x + p_1 A_1 - p_2 A_2 = \dot{m} V_2 \cos \alpha - \dot{m} V_1$$
$$= 0 \; (\text{since } V_2 \cos \alpha = V_1)$$

and

$$S_y = \dot{m} V_2 \sin \alpha.$$

Note that in applying the momentum equation in the x-direction, the area acted on by the pressures p_1 and p_2 is A_1 in both cases and not A_2 for p_2. Also, the fluid–guidevane interaction force S (with components S_x and S_y) takes into account the forces acting on both sides of the guidevane.

The components of the net force exerted by the fluid on a single guidevane are thus

$$L = S_y = \dot{m} V_2 \sin \alpha = \dot{m} V_1 \tan \alpha$$

and

$$D = S_x = (p_1 - p_2) A_1$$
$$= \tfrac{1}{2} \rho V_1^2 \tan^2 \alpha \, A_1$$
$$= \tfrac{1}{2} \dot{m} V_1 \tan^2 \alpha.$$

We have introduced the symbols L and D here because the force exerted by the fluid on a guidevane normal to the approach-flow direction is the lift L and the corresponding force in the x-direction is the associated drag D. The *drag associated with lift* is termed the *induced drag* and we see that

$$D = \tfrac{1}{2} L \tan \alpha$$

a result which holds good in much more general situations. We also note that another general aspect of aerodynamic lift is that it is invariably achieved through a change in momentum flowrate due to a change in flow direction.

A rather sophisticated approach to the calculation of aerofoil lift is through the concept of *circulation*. If we draw a closed loop around an aerofoil, split the loop into many infinitesimal segments each of length δs, then the circulation is just the sum of the product of δs and the tangential component of velocity corresponding to each segment. For one of our guidevanes we choose the loop shown in Figure 10.9(c) with identical curved segments in adjacent channels separated by a distance equal to the pitch *w*. In this case, the circulation Γ (gamma) is simply

$$\Gamma = V_2 \sin \alpha \, w$$

because over section ① the velocity V_1 is normal to the loop and so the contribution to Γ is zero, along the curved boundaries of the loop the sum of all the contributions along the upper boundary exactly cancel those along the lower boundary, and all that is left is $V_2 \sin \alpha$ along section ② of length *w*.

According to what is called the Kutta–Joukowski theorem, the lift per unit span is given by

$$\rho V_1 \Gamma$$

so that for the guidevane

$$L = \rho V_1 \Gamma s$$
$$= \rho V_1 V_2 \sin \alpha \, w s$$
$$= \dot{m} V_2 \sin \alpha = \dot{m} V_1 \tan \alpha$$

which is exactly the same result that we obtained from the momentum equation. Unfortunately it is usually far more difficult to calculate the circulation for an isolated aerofoil than it was for the simple guidevane considered here.

EXAMPLE E10.8
Hot gas with a density of 0.2 kg/m³ and velocity of 200 m/s leaves a combustion chamber and is deflected through an angle of 20° by a set of curved guidevanes. If the approach-flow cross-sectional area is 0.5 m², calculate the components of force exerted on the guidevane set in directions parallel to and perpendicular to the approach-flow direction. If the gas

exhausts from the guidevanes at atmospheric pressure (1.01 bar), calculate the stagnation pressure of the flow and the pressure drop across the set of guidevanes.

SOLUTION

$\rho = 0.2$ kg/m^3; $V_1 = 200$ m/s; $\alpha = 20°$; $A_1 = 0.5$ m^2; $p_2 = 1.01 \times 10^5$ Pa

We start by calculating the mass flowrate

$$\dot{m} = \rho A_1 V_1 = 0.2 \times 0.5 \times 200 = 20 \text{ kg/s.}$$

The total lift force (i.e. the component of force perpendicular to the approach-flow direction) is thus

$$L = \dot{m} V_1 \tan \alpha = 20 \times 200 \times \tan 20° = 1456 \text{ N}$$

and the total drag force

$$D = \tfrac{1}{2} L \tan \alpha = 0.5 \times 1456 \times \tan 20° = 265 \text{ N.}$$

For the pressure drop across the guidevanes we have

$$p_1 - p_2 = \tfrac{1}{2} \rho (V_2^2 - V_1^2)$$

and so we need to calculate the velocity at outlet V_2.
From the continuity equation we have

$$V_2 = \frac{\dot{m}}{\rho A_1 \cos \alpha} = \frac{20}{0.2 \times 0.5 \times \cos 20°} = 213 \text{ m/s}$$

and so

$$p_1 - p_2 = 0.5 \times 0.2(213^2 - 200^2) = 530 \text{ Pa.}$$

The stagnation pressure can be evaluated from the values of p_2 and V_2, i.e.

$$p_0 = (1.01 \times 10^5) + (0.5 \times 0.2 \times 213^2) = 1.055 \times 10^5 \text{ Pa} \quad \text{or} \quad 1.055 \text{ bar.}$$

10.10 Jet impinging on a flat surface

In contrast to the internal flows we have considered so far, there are many situations in which the momentum flowrate of a free jet, usually liquid, is reduced in one direction and increased in the perpendicular direction because the jet is deflected by impingement on a stationary or moving object. We consider first the case of deflection by a flat stationary plate held at an angle to the jet, and secondly the effect of moving the plate. Also, we shall restrict ourselves to a liquid jet discharging into the atmosphere.

10.10.1 *Stationary plate*

If the jet shown in Figure 10.11(a) has a circular cross-section and the plate is held normal to it (i.e. $\alpha = 0$)°, the deflected fluid will flow radially outwards over the plate. The net momentum outflow from the fluid control volume (Figure 10.11(b)) is then almost zero and there is no difficulty on applying the momentum equation. The situation becomes much more complicated if the plate is held at an angle, as shown in Figure 10.11(a), and is unsuited to the simple analysis presented in this chapter and the previous one. To simplify matters we consider only flat jets, i.e. we regard what we see in Figure 10.11(a) as a section through a sheet of fluid leaving a rectangular nozzle of height t and width w with $w \gg t$.

Before proceeding further, we need to make a number of assumptions. First, there is no mixing between the liquid jet and the surrounding air. Second, the flow is entirely frictionless. Third, the pressure acting on the free surface of the liquid is equal to the ambient pressure B.

The frictionless-flow assumption has two important consequences: first, Bernoulli's equation can be applied to the flow and, second, the fluid–interaction force S must act perpendicular to the plate surface because the liquid pressure is the only stress exerted by the liquid on the plate surface.

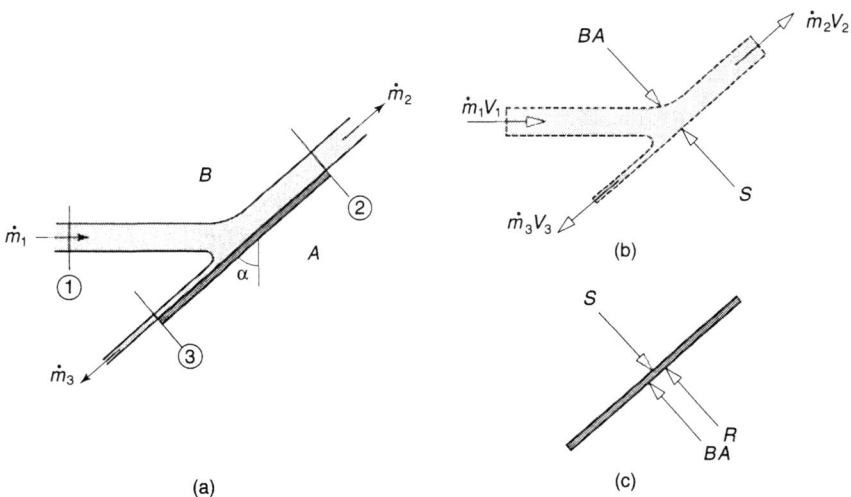

Figure 10.11 *Liquid jet impinging on a stationary flat plate: (a) flow geometry; (b) fluid control volume; (c) forces exerted*

The third assumption also means that the pressures at sections ①, ② and ③ are all equal to the ambient pressure. According to Bernoulli's equation, therefore,

$$p_0 = B + \tfrac{1}{2} \rho V_1^2 = B + \tfrac{1}{2} \rho V_2^2 = B + \tfrac{1}{2} \rho V_3^2$$

from which we deduce

$$V_1 = V_2 = V_3.$$

From the continuity equation we have

$$\dot{m}_1 = \rho A_1 V_1, \quad \dot{m}_2 = \rho A_2 V_2 \quad \text{and} \quad \dot{m}_3 = \rho A_3 V_3$$

while from the overall mass-conservation equation

$$\dot{m}_1 = \dot{m}_2 + \dot{m}_3.$$

We conclude therefore that

$$A_1 = A_2 + A_3.$$

Because the fluid–structure interaction force S acts normal to the plate, it is convenient to apply the momentum equation in directions normal and parallel to the plate. Normal to the plate we find

$$-S + BA = -\dot{m}_1 V_1 \cos \alpha$$

and from the condition for static equilibrium of the plate (see Figure 10.10(c)), we have

$$S - R - BA = 0$$

so that the hydrodynamic reaction force R is given by

$$R = \dot{m}_1 V_1 \cos \alpha.$$

The momentum equation taken parallel to the plate gives

$$0 = \dot{m}_2 V_2 - \dot{m}_3 V_3 - \dot{m}_1 V_1 \sin \alpha.$$

If we substitute for the mass flowrates from the continuity equation and also use the results $V_1 = V_2 = V_3$ and $A_1 = A_2 + A_3$, we find

$$A_2 = \tfrac{1}{2} A_1 (1 + \sin \alpha) \quad \text{and} \quad A_3 = \tfrac{1}{2} A_1 (1 - \sin \alpha)$$

so that, due to the impingement on the plate, the initial flowrate $\dot{m}_1$ is split in the ratio

$$\frac{\dot{m}_2}{\dot{m}_3} = \frac{1 + \sin \alpha}{1 - \sin \alpha}.$$

10.10.2 Moving plate

If the plate in the previous section is moving at velocity V_P in the same direction as the jet, we have an unsteady-flow problem relative to the surroundings. To convert this to a steady-flow problem we impose a velocity $-V_P$ on both the plate and the jet. The relative velocity between the jet and the plate is then $(V_1 - V_P)$ and, with $(V_1 - V_P)$ replacing V_1, the analysis of section 10.10.1 remains valid.

EXAMPLE 10.9

A water jet with a velocity of 10 m/s has a rectangular cross-section of 25 mm by 100 mm. Calculate the hydrodynamic reaction force exerted on a plate held at 60° to the jet (a) if the plate is stationary and (b) if the plate is moving in the same direction as the jet at a velocity of 5 m/s. Calculate also the ratio in which the approach flow is split by the plate.

SOLUTION

(a) $V_P = 0$; $\rho = 1000 \text{ kg/m}^3$; $V_1 = 10 \text{ m/s}$; $A_1 = 0.025 \times 0.1 \ m^2$; $\alpha = 30°$

We calculate first the mass flowrate of the jet

$$\dot{m}_1 = \rho A_1 V_1 = 1000 \times 0.025 \times 0.1 \times 10 = 25 \text{ kg/s}.$$

From section 10.10.1 we have

$$R = \dot{m}_1 V_1 \cos \alpha = 25 \times 10 \times \cos 30° = 217 \text{ N}$$

and the flow is split in the ratio

$$\frac{\dot{m}_2}{\dot{m}_3} = \frac{1 + \sin \alpha}{1 - \sin \alpha} = 3 : 1.$$

(b) $V_P = 5 \text{ m/s}$

As we pointed out in section 10.10.2, the essential change is that V_1 is now replaced by $V_1 - V_P$, i.e. by 5 m/s. Then

$$\dot{m}_1 = 1000 \times 0.025 \times 0.1 \times 5 = 12.5 \text{ kg/s}$$

and

$$R = 12.5 \times 5 \times \cos 30° = 54.1 \text{ N}.$$

10.11 Pelton impulse hydraulic turbine

An impulse hydraulic turbine is one in which a high-speed jet of water impinges in turn on each of a series of specially shaped guidevanes, usually called buckets, attached to the rim of a rotating wheel or runner as shown in Figure 1.9. The classic bucket shape shown in Figure 1.9 is due to the American engineer Lester Allen Pelton (1829–1908). The Pelton bucket is designed to split the water jet into two, turn each half through about 165° and eject the water to either side of the runner. The symmetrical design ensures that there is no axial load on the runner and the high deflection angle produces a momentum change within a few per cent of the maximum possible. The diameter of the runner of a large Pelton turbine is about 4 m with typically 20–24 buckets equally spaced around its rim. Water is supplied to the turbine through one or more nozzles (2, 4 or 6 are common choices) with spear (or needle) valves and deflectors to control the flowrate. The high pressure required to create the high-speed water jet through a nozzle is achieved by situating the turbine at the end of a supply pipe (called a penstock) fed from a reservoir often more than 1000 m above the turbine.

In the following simple analysis we neglect the influence of rotation on the jet–bucket interaction and assume that the jet impinges on a Pelton bucket moving at a linear velocity V_B (see Figure 10.12(a)) equal to the peripheral velocity of a bucket at pitch radius R on a runner rotating at N rps, i.e.

$$V_B = 2\pi NR.$$

As we did for a jet impinging on a moving flat surface (section 10.10.2), we consider the water flow relative to the Pelton bucket. If we apply the linear momentum equation to the fluid control volume shown in Figure 10.12(b), we have

$$-S + BA_B = -\dot{m}_1 V_1 \cos \alpha - \dot{m}_1 V_1$$

where A_B is the projected area of the bucket and V_1 is the velocity of the water jet relative to the bucket, i.e.

$$V_1 = V_J - V_B$$

V_J being the actual velocity of the jet produced by the nozzle. The magnitude of the relative water velocity is the same at the control volume inlet and outlet because we have again assumed that the pressure on the surface of the water is

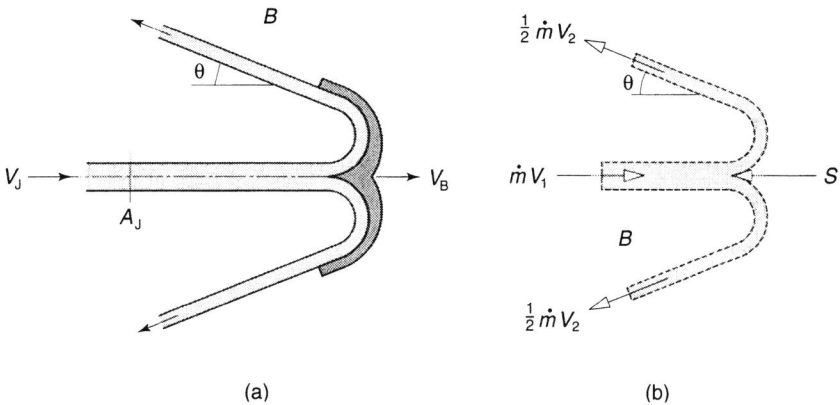

Figure 10.12 *Pelton-wheel bucket: (a) flow configuration; (b) flow control volume*

constant (at the ambient pressure) and that there is negligible friction between the bucket surface and the water flowing over it.

Although the actual mass flowrate of the jet leaving the nozzle is given by

$$\dot{m}_J = \rho A_J V_J,$$

the effective mass flowrate entering the control volume $\dot{m}_1$ must correspond to the relative velocity so that

$$\dot{m}_1 = \rho A_J V_1 = \rho A_J (V_J - V_B).$$

The net force F_B exerted by the water on the Pelton bucket is then given by

$$F_B = S - BA_B = \rho A_J (V_J - V_B)^2 (1 + \cos \alpha).$$

Some care is necessary in calculating the power generated by the turbine due to the force F_B acting on the runner. The reason for this is that as one bucket sweeps into the path of the water jet, a substantial amount of water is still being deflected by the previous bucket and for a period of time at least two buckets will be generating power simultaneously. We take this effect into account for a single jet as follows.

Since the time for one complete turn of the runner is $1/N$, if there are n_B buckets in total, the time for which the jet is intercepted by any one bucket during one turn must be

$$t_i = \frac{1}{n_B N}.$$

During this time the total mass of fluid leaving the nozzle is

$$m = \dot{m}_J t_i = \rho A_J V_J t_i.$$

However, since the mass flowrate through the fluid control volume is only $\dot{m}_1$, the time the mass of fluid m will be in contact with the bucket, and so generating power, must be

$$t_c = \frac{m}{\dot{m}_1} = \frac{m}{\rho A_J V_1} = \frac{V_J}{V_1} t_i.$$

The angular displacement of the runner during time t_c is

$$2\pi N t_c$$

and the work done by the torque $F_B R$ due to the force F_B acting at a radius R is

$$F_B R.2\pi N t_c.$$

Since there are n_B buckets, the total work for one turn of the runner is

$$n_B F_B R.2\pi N t_c$$

and the corresponding power generated by the turbine must be given by

$$P = \frac{n_B F_B R.2\pi N t_c}{n_B t_i} = \frac{F_B R.2\pi N t_c}{t_i}$$

i.e. the actual power is greater than might have been expected by the factor t_c/t_i.

Since

$$\frac{t_c}{t_i} = \frac{V_J}{V_1} = \frac{V_J}{V_J - V_B} \quad \text{and} \quad V_B = 2\pi N R$$

and also

$$F_B = \rho A_J (V_J - V_B)^2 (1 + \cos\alpha),$$

we see that

$$P = \rho V_J A_J V_B (V_J - V_B)(1 + \cos\alpha)$$
$$= \dot{m}_J V_B (V_J - V_B)(1 + \cos\alpha).$$

This result applies for a Pelton turbine supplied by a single nozzle. For a turbine with n_J jets, the power output is $n_J P$. As a final point, we can see from the above result that the power is a maximum for a given jet speed V_J when the peripheral bucket speed V_B is equal to $\frac{1}{2} V_J$.

EXAMPLE 10.10
A Pelton turbine with a runner diameter (i.e. bucket pitch circle) of 2.85 m rotates at a speed of 8.3 rps. Water is supplied through six nozzles each producing a jet 0.17 m in diameter with a speed of 157 m/s. If the Pelton buckets deflect the water through 165°, calculate the power generated by the turbine.

SOLUTION
$R = 1.425$ m; $\quad \rho = 1000$ kg/m³; $\quad n_J = 6$; $\quad D_J = 0.17$ m;
$V_J = 157$ m/s; $\quad N = 8.3$ rps; $\quad \alpha = 15°$

We start by calculating A_J, $\dot{m}_J$ and V_B as follows

$$A_J = \frac{\pi D_J^2}{4} = 0.0227 \text{ m}^2$$

$$\dot{m}_J = \rho A_J V_J = 3564 \text{ kg/s}$$

$$V_B = 2\pi N R = 74.3 \text{ m/s}.$$

The power corresponding to each jet is then

$$P = \dot{m}_J V_B (V_J - V_B)(1 + \cos \alpha) = 4.3 \times 10^7 \text{ W} \quad \text{or} \quad 43 \text{ MW}$$

and the total power for six jets is 258 MW.

Comment: The data used for this example are close to those for two vertical Pelton turbine sets, among the most powerful ever built, installed in Sellrain-Silz, Austria.

10.12 Summary

In this chapter we have shown how to apply the linear momentum equation, together with the continuity equation and either Bernoulli's equation or some other information about pressure changes, to the analysis of a diverse range of problems. A key aim was to demonstrate that we have established a relatively simple theoretical basis which can give quite accurate and useful information about the performance of such complex machines as jet and rocket engines, the jet pump, and the Pelton turbine. Other examples included pipe bends and junctions and a cascade of guidevanes.

For a given problem, the student should be able to

- identify a fluid control volume suitable for the application of the linear momentum equation
- apply the linear momentum equation to the flow through the control volume, either in one direction or two perpendicular directions
- apply the continuity equation to the flow through the control volume and also, for flows with multiple inlets and/or outlets, the overall mass conservation equation
- use either Bernoulli's equation or other information to determine pressure changes throughout the flow
- where appropriate, calculate the hydrodynamic forces created by fluid passing through or around a component or machine.

10.13 Self-assessment problems

10.1   The jet engine on a test stand as shown in Figure P10.1 takes in air at atmospheric conditions (20°C and 1.01 bar) at section ①, where $V_1 = 200$ m/s and $A_1 = 0.3$ m². The fuel : air ratio is 1 : 40. The exhaust gases leave at atmospheric pressure at section ②, where $V_2 = 1000$ m/s and $A_2 = 0.25$ m². Compute the test-stand reaction R which balances the thrust of the engine.

   Suppose that a deflector is attached to the exit of the jet engine, as shown at the right of the figure. What will the reaction R on the test stand be now? How does this reaction relate to the braking ability of the deflector?

   (Answers: 59.4 kN, −66.6 kN)

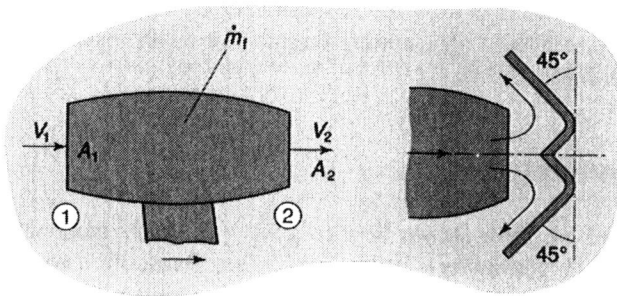

Figure P10.1

10.2   (a) A liquid atomiser has the configuration shown in Figure P10.2. The liquid is accelerated through the nozzle and impinges on a cone attached to the nozzle by a thin rod. The nozzle is circular in cross-section and

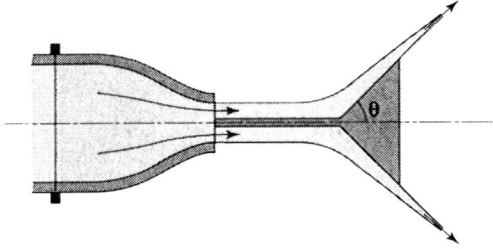

Figure P10.2

coaxial with the rod and cone. Show that the net hydrodynamic force R to be withstood by the flange bolts is

$$R = \frac{\rho \dot{Q}^2}{2A_2} \left(\frac{A_1}{A_2} + \frac{A_2}{A_1} - 2 \cos \theta \right)$$

where ρ is the liquid density, $\dot{Q}$ the liquid volume flowrate, A_1 the upstream area of the nozzle, A_2 the nozzle exit area and θ is the half angle of the cone. Assume that, external to the nozzle, the liquid pressure is equal to that of its surroundings, that there are no losses and that gravitational effects are negligible, as is the influence of the thin rod on the flow.

(b) A liquid of density 900 kg/m^3 flows through a nozzle-cone atomiser as in Figure P10.2. The nozzle has a circular upstream cross-section 50 mm in diameter and an exit diameter of 15 mm. If the liquid mass flowrate is 1.35 kg/s and the ambient pressure 1.01 bar, calculate the stagnation pressure of the liquid and the separate forces on the nozzle and the cone. The cone half angle is 60°.
(Answers: 1.33 bar, 52.7 N, 5.73 N)

10.3 (a) Water flows into a hydraulic turbine through a horizontal pipe of cross-sectional area A and then over a hub of cross-sectional area CA, where C is a numerical constant less than 1. The arrangement is shown in Figure P10.3. If V is the velocity of the water in the pipe and ρ its density, show that the water exerts a force on the hub given by

$$R = \tfrac{1}{2} \rho A \left(\frac{CV}{1-C} \right)^2$$

and find the direction of R (upstream or downstream). Assume that the flow is steady and frictionless, that gravitational effects are negligible, and that the pressure and velocity are uniformly distributed across sections ① and ②. Note that the hub is connected rigidly to the shaft

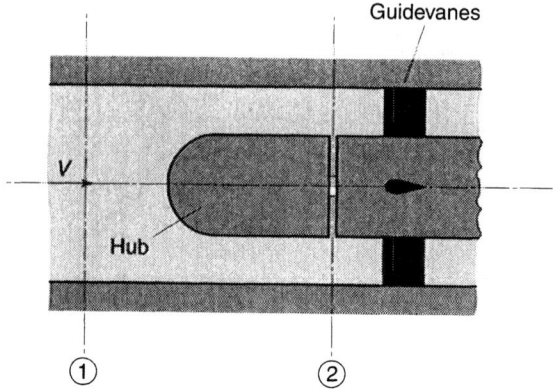

Figure P10.3

supporting the guide vanes by a spindle of negligible cross-section in such a way that the pressure at section ② is constant across the back of the hub and equal to the value in the water stream at section ②.

(b) Calculate R for a turbine with an inlet pipe 10 m in diameter and a hub 8.5 m in diameter for a water flowrate of 2×10^5 kg/s. Calculate also the pressure difference between the forward stagnation point on the nose of the hub and the rear of the hub.

(Answers: 1.73 MN, 0.42 bar)

10.4 (a) The rate of flow of a liquid of density ρ through a duct of square cross-section is controlled by a hinged plate, as shown in Figure P10.4. Show that when the plate is at an angle θ as shown, the force F acting on the hinge is given by

$$F = (p_0 - p_E) A \cos \theta$$

where p_0 is the upstream liquid pressure when the duct is completely blocked by the plate (i.e. $\theta = 0°$), p_E is the liquid pressure immediately downstream of the plate and A is the area of both the duct cross-section

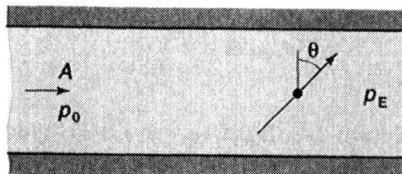

Figure P10.4

and the plate. Assume that the flow is steady, one dimensional, frictionless and incompressible.

(b) For a duct of cross-sectional area 4 m^2 the mass flowrate is found to be 1000 kg/s when the opening angle of the plate is 30°. Calculate the upstream flow velocity and the force F if the liquid density is 800 kg/m^3.

(Answers: 0.313 m/s, 7539 N)

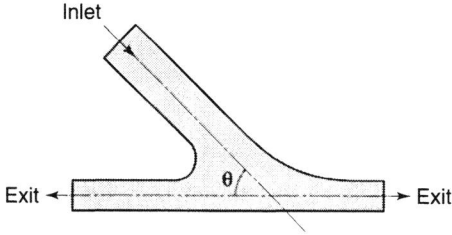

Inlet

Exit

θ

Exit

Figure P10.5

10.5 (a) A liquid stream is split and deflected by the pipe junction shown in Figure P10.5 and discharged to surroundings at ambient pressure. Gravitational effects may be neglected and the flow may be assumed to be steady, one dimensional, incompressible and frictionless. Each of the two exits has an area one-quarter that of the inlet area A. Show that the external reaction force R required to hold the pipe junction in place is given by

$$R = \frac{5\dot{m}^2}{2\rho A}$$

where $\dot{m}$ is the mass flowrate of the liquid and ρ its density. In which direction must R act? [*Hint:* Show that the outflow conditions are the same for each exit and use this information to guide the solution.]

(b) In a particular case, the liquid is crude oil of relative density 0.85 and the inlet a circular pipe of diameter 1 m. The static pressure of the oil at inlet to the pipe junction is 1.04 bar and the ambient pressure is 1.01 bar. Calculate the oil mass flowrate, the stagnation pressure p_0 and the reaction force R.

(Answers: 1024 kg/s, 1.05 bar, 3.93 kN)

10.6 (a) Figure P10.6 shows the design of a simple jet pump in which a secondary airflow is sucked through a tube of large diameter D by air flowing at high speed through a central tube of diameter d. If the mass

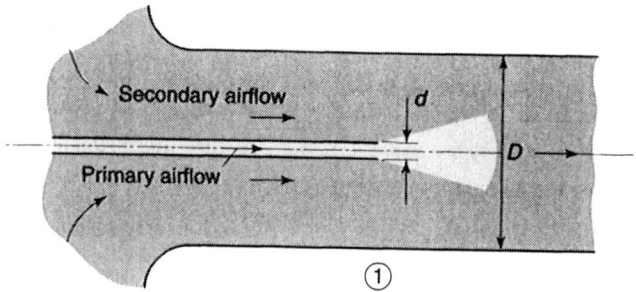

Figure P10.6

flowrate of air through the central tube is $\dot{m}_N$, show that the secondary mass flowrate $\dot{m}_S$ is given by the quadratic equation

$$(\dot{m}_N + \dot{m}_S)^2 - \frac{\dot{m}_N^2}{x} - \frac{\dot{m}_S^2}{1-x} + \frac{\dot{m}_S^2}{2(1-x)^2} = 0$$

where $x = (d/D)^2$. Assume that the flow is one dimensional and incompressible throughout, that up to plane ① (see Figure P10.6) both flows are loss-free (i.e. frictionless) and that the static pressure is the same in both flows at plane ①. The secondary flow is drawn from the surrounding atmosphere and the total discharge is to the atmosphere. Note that Bernoulli's equation does *not* apply downstream of plane ①. [*Hint:* write equations for both mass flowrates, the stagnation pressure of the secondary flow and apply the momentum equation to the flow downstream of plane ①.]

(b) A jet pump is designed with $D = 100$ mm and $d = 10$ mm. Calculate the ratio between the two mass flowrates, $\dot{m}_S/\dot{m}_N$. If the air velocity in the central tube is 150 m/s, calculate the secondary flow airspeed, the static pressure at plane ① and the stagnation pressure of the airflow in the central tube. The air density may be taken as 1.2 kg/m³ and the atmospheric pressure as 1.01 bar.

(Answers: 12.2, 18.5 m/s, 1.0079 bar, 1.143 bar)

Physical properties of selected gases and liquids and other data

For convenience in some of the tables below we have used the exponent notation, i.e. 1.787 E−3 is equivalent to 1.787×10^{-3}, etc. The data in Tables A.1 to A.5 have been compiled primarily from Kaye and Laby (1973) and White (1994).

In Tables A.2 to A.5 and A.8, the symbols have the following meanings

c	speed of sound
M	molecular weight
p	pressure
p_V	vapour pressure
R	specific gas constant
T	temperature
γ	ratio of specific heats
μ	dynamic viscosity
ν	kinematic viscosity
ρ	density
σ	surface tension

Table A.1 *Some universal constants*

Universal gas constant	$\Re = 8314.510$ J/kmol.K (= MR)
Standard atmosphere	1 atm = 1.01325 E+5 Pa
Speed of light in vacuum	$c_0 = 2.9979$ E+8 m/s
Standard acceleration due to gravity	g = 9.80665 m/s^2

Table A.2 *Physical properties of water at 1 atm*

T (°C)	ρ (kg/m³)	μ (Pa.s)	ν (m²/s)	σ (N/m)	p_v (Pa)	c (m/s)
0	999.8	1.787 E−3	1.787 E−6	7.57 E−2	6.107 E+2	1402
4	1000.0	1.573 E−3	1.573 E−6	7.49 E−2	8.130 E+2	1422
10	999.7	1.304 E−3	1.307 E−6	7.42 E−2	1.228 E+3	1447
20	998.2	1.002 E−3	1.004 E−6	7.28 E−2	2.338 E+3	1482
30	995.7	7.975 E−4	8.009 E−7	7.12 E−2	4.245 E+3	1509
40	992.2	6.540 E−4	6.580 E−7	6.96 E−2	7.381 E+3	1529
50	988.0	5.477 E−4	5.534 E−7	6.79 E−2	1.235 E+4	1543
60	983.2	4.674 E−4	4.745 E−7	6.62 E−2	1.993 E+4	1551
70	977.8	4.048 E−4	4.134 E−7	6.44 E−2	3.117 E+4	1555
80	971.8	3.554 E−4	3.650 E−7	6.26 E−2	4.738 E+4	1555
90	965.3	3.155 E−4	3.260 E−7	6.08 E−2	7.012 E+4	1550
100	958.4	2.829 E−4	2.940 E−7	5.88 E−2	1.013 E+5	1542
110	950.6				1.43 E+5	
120	942.5				1.99 E+5	
200	865.1				1.555 E+6	
300	712.3				8.593 E+6	

Table A.3 *Physical properties of air at 1 atm*

T (°C)	ρ (kg/m^3)	μ (Pa.s)	ν (m^2/s)	c (m/s)
−40	1.514	1.57 E−5	1.04 E−5	306.2
−20	1.395	1.63 E−5	1.17 E−5	319.1
0	1.292	1.71 E−5	1.32 E−5	331.4
5	1.269	1.73 E−5	1.36 E−5	334.4
10	1.247	1.76 E−5	1.41 E−5	337.4
15	1.225	1.80 E−5	1.47 E−5	340.4
20	1.204	1.82 E−5	1.51 E−5	343.4
25	1.184	1.85 E−5	1.56 E−5	346.3
30	1.165	1.86 E−5	1.60 E−5	349.1
40	1.127	1.87 E−5	1.66 E−5	354.7
50	1.109	1.95 E−5	1.76 E−5	360.3
60	1.060	1.97 E−5	1.86 E−5	365.7
70	1.029	2.03 E−5	1.97 E−5	371.2
80	0.9996	2.07 E−5	2.07 E−5	376.6
90	0.9721	2.14 E−5	2.20 E−5	381.7
100	0.9461	2.17 E−5	2.29 E−5	386.9
150	0.8343	2.38 E−5	2.85 E−5	412.3
200	0.7461	2.53 E−5	3.39 E−5	434.5
250	0.6748	2.75 E−5	4.08 E−5	458.4
300	0.6159	2.98 E−5	4.84 E−5	476.3
400	0.5243	3.32 E−5	6.34 E−5	514.1
500	0.4565	3.64 E−5	7.97 E−5	548.8
1000	0.2772	5.04 E−5	1.82 E−4	694.8

Table A.4 *Physical properties of common liquids at 20°C and 1 atm*

Liquid	M	ρ (kg/m³)	μ (Pa.s)	σ (N/m)	ρ_v (Pa)	K (Pa)	c (m/s)
Benzene C_6H_6	78.1	879	6.47 E-4	2.89 E-2 A	1.01 E+4	1.4 E+9	1320
Carbon tetrachloride CCl_4	153.8	1632	9.72 E-4	2.70 E-2 V	1.20 E+4	9.65 E+8	940
Castor oil		950	9.86 E-1				1900
Ethanol C_2H_5OH	46.1	789	1.20 E-3	2.28 E-2 V	5.7 E+3	9.0 E+8	1162
Ethylene glycol $C_2H_6O_2$	62.1	1110	1.99 E-2		5.7 E+3	9.0 E+8	1162
Freon 12 CCl_2F_2	120.9	1327	2.62 E-4				
Glycerol $C_3H_8O_3$	92.1	1260	1.49	6.34 E-2 A	1.4 E-2	4.34 E+9	1860
Mercury Hg	200.6	13546	1.56 E-3	4.72 E-1 V	1.1 E-3	2.55 E+10	1454
Methanol CH_3OH	32.0	791	5.94 E-4	2.26 E-2 A	1.34 E+4	8.3 E+8	1121
Olive oil		900	8.4 E-2			1.60 E+9	1440
Paraffin oil		804	1.92 E-3	2.8 E-2	3.11 E+3	1.6 E+9	1315
Petrol		680	2.92 E-4	2.16 E-2	5.51 E+4	9.58 E+8	
SAE 10 W oil		870	1.04 E-1	3.6 E-2		1.31 E+9	
SAE 30 W oil		891	2.9 E-1	3.5 E-2		1.38 E+9	
SAE 50 W oil		902	8.6 E-1				
Sea water (3.5% salinity)		1025	1.07 E-3	7.28 E-2	2.34 E+3	2.33 E+9	1522
Water	18.0	998.2	1.00 E-3	7.28 E-2	2.34 E+3	2.19 E+9	1482

For surface tension, A = against air, V = against own vapour.
There is a significant decrease in the dynamic viscosity of liquids with increase in temperature (see Figure 2.3). Below 100 bar, the viscosity of a liquid is practically independent of pressure.

Table A.5 *Physical properties of common gases at 20°C and 1 atm*

Gas	M	R $(m^2/s^2.K)$	ρ (kg/m^3)	μ[a] $(Pa.s)$	c[b] (m/s)	γ
Dry air	28.97	287.0	1.205	1.82 E–5	343	1.401
Argon Ar	39.95	208.1	1.662	2.23 E–5	319	1.667
Butane C_4H_{10}	58.12	143.0				
Carbon dioxide CO_2	44.01	188.9	1.830	1.47 E–5	268	1.300
Carbon monoxide CO	28.01	296.8	1.165	1.75 E–5	349	1.401
Chlorine Cl_2	70.91	117.3	2.956	1.32 E–5	219	1.395
Ethane C_2H_6	30.07	276.5	1.251		310*	1.22
Ethylene C_2H_4	28.05	296.4	1.167	1.03 E–5	331	1.264
Helium He	4.003	2077	0.1664	1.96 E–5	1007*	1.63
Hydrogen H_2	2.016	4124	0.0838	8.84 E–6	1332*	1.407
Methane CH_4	16.04	518.4	0.66	1.10 E–5	446*	1.313
Neon Ne	20.18	412.0	0.839	3.13 E–5	450	1.642
Nitrogen N_2	28.01	296.8	1.165	1.76 E–5	349	1.401
Nitric oxide NO	30.01	277.1	1.248	1.90 E–5	336	1.394
Nitrous oxide N_2O	44.02	188.9	1.830	1.46 E–5	266*	1.324
Oxygen O_2	32.00	259.8	1.330	2.04 E–5	326	1.400
Sulphur dioxide SO_2	64.07	129.8	2.664	1.26 E–5	219	1.26
Water vapour H_2O	18.0	461.4	0.7498	9.7 E–6	425	1.334

[a] The dynamic viscosity of gases increases by between 18 and 28% over the temperature range 20 to 100°C (see Figure 2.3). Below 10 bar the viscosity of a gas is practically independent of pressure.
[b] The speed of sound for an ideal gas can be calculated from $c = \sqrt{(\gamma RT)}$ where T is the absolute temperature (K) of the gas. Cases where this calculated value differs slightly from the accepted value of c have been indicated by an asterisk (*).

TABLE A.6 *Physical quantities, symbols, SI units and dimensions*

Quantity	Symbol(s)	SI unit(s)	Dimensions
Plane angle	θ	rad	1
Mass	m	kg	M
Length	ℓ, x, y, z, R, D	m	L
Area	a, A	m^2	L^2
Volume	ϑ	m^3	L^3
Time	t	s	T
Frequency	f	s^{-1} = Hz	1/T
Angular velocity	ω	rad/s	1/T
Mass flow rate	$\dot{m}$	kg/s	M/T
Velocity or speed	V	m/s	L/T
Acceleration	a	m/s^2	L/T^2
Acceleration due to gravity	g	m/s^2	L/T^2
Volumetric flowrate	$\dot{Q}$	m^3/s	L^3/T
Momentum flowrate	$\dot{M}$	kg.m/s	ML/T
Density	ρ	kg/m^3	M/L^3
Force	D, F, R, S	N = kg.m/s^2	ML/T^2
Pressure	p	Pa = N/m^2 (also bar = 10^5 Pa)	M/LT^2
Shear stress	τ	Pa	M/LT^2
Surface tension	σ	N/m	M/T^2
Dynamic viscosity	μ	Pa.s = kg/m.s	M/LT
Kinematic viscosity	ν	m^2/s	L^2/T
Work	W	J = N.m	ML^2/T^2
Energy	E	J	ML^2/T^2
Power	P	W = J/s	ML^2/T^3

TABLE A.7 *SI prefixes*

Factor	Prefix	Symbol	Factor	Prefix	Symbol
10^9	giga	G	10^{-3}	milli	m
10^6	mega	M	10^{-6}	micro	μ
10^3	kilo	k	10^{-9}	nano	n

Table A.8 *Physical properties of the Standard Atmosphere*

z' (m)	T (K)	p (Pa)	ρ (kg/m³)	z' (m)	T (K)	p (Pa)	ρ (kg/m³)
−500	291.41	107 508	1.2854	12 500	216.66	17 847	0.2870
0	288.16	101 350	1.2255	13 000	216.66	16 494	0.2652
500	284.91	95 480	1.1677	13 500	216.66	15 243	0.2451
1 000	281.66	89 889	1.1120	14 000	216.66	14 087	0.2265
1 500	278.41	84 565	1.0583	14 500	216.66	13 018	0.2094
2 000	275.16	79 500	1.0067	15 000	216.66	12 031	0.1935
2 500	271.91	74 684	0.9570	15 500	216.66	11 118	0.1788
3 000	268.66	70 107	0.9092	16 000	216.66	10 275	0.1652
3 500	265.41	65 759	0.8633	16 500	216.66	9 496	0.1527
4 000	262.16	61 633	0.8191	17 000	216.66	8 775	0.1411
4 500	258.91	57 718	0.7768	17 500	216.66	8 110	0.1304
5 000	255.66	54 008	0.7361	18 000	216.66	7 495	0.1205
5 500	252.41	50 493	0.6970	18 500	216.66	6 926	0.1114
6 000	249.16	47 166	0.6596	19 000	216.66	6 401	0.1029
6 500	245.91	44 018	0.6237	19 500	216.66	5 915	0.0951
7 000	242.66	41 043	0.5893	20 000	216.66	5 467	0.0879
7 500	239.41	38 233	0.5564	22 000	218.6	4 048	0.0645
8 000	236.16	35 581	0.5250	24 000	220.6	2 972	0.0469
8 500	232.91	33 080	0.4949	26 000	222.5	2 189	0.0343
9 000	229.66	30 723	0.4661	28 000	224.5	1 616	0.0251
9 500	226.41	28 504	0.4387	30 000	226.5	1 197	0.0184
10 000	223.16	26 416	0.4125	40 000	250.4	287	0.0040
10 500	219.91	24 455	0.3875	50 000	270.7	80	0.0010
11 000	216.66	22 612	0.3637	60 000	255.7	22	0.0003
11 500	216.66	20 897	0.3361	70 000	219.7	6	0.0001
12 000	216.66	19 312	0.3106				

Table A.9 *Areas, centroids and second moments of area for common shapes (❂ denotes centroid and C – C axis of rotation for I_C)*

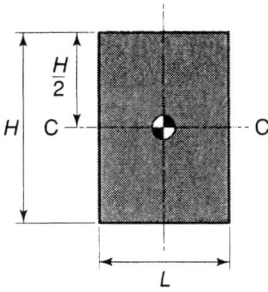

Rectangle:

$A = LH$

$I_C = \dfrac{AH^2}{12} = \dfrac{LH^3}{12}$

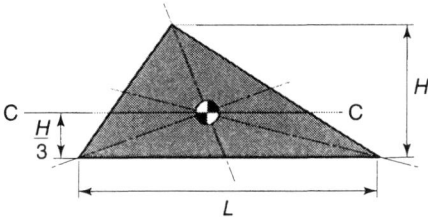

Triangle:

$A = \dfrac{LH}{2}$

$I_C = \dfrac{AH^2}{18} = \dfrac{LH^3}{36}$

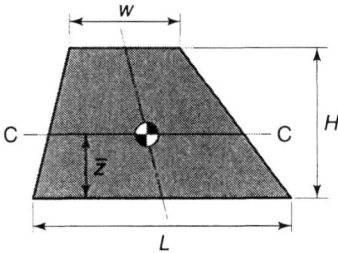

Trapezium:

$A = \dfrac{(w+L)H}{2}$

$\bar{z} = \dfrac{(2w+L)H}{3(w+L)}$

$I_C = \dfrac{(w^2+4wL+L^2)H^3}{36(w+L)}$

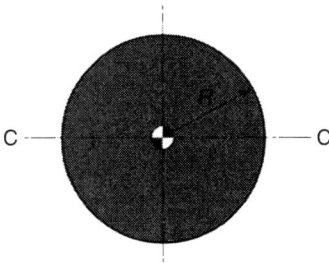

Circle:

$A = \pi R^2$

$I_C = \dfrac{AR^2}{4} = \dfrac{\pi R^4}{4}$

(Continued)

Table A.9 *(continued)*

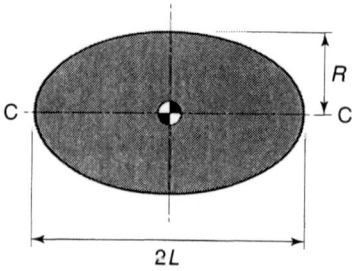

Ellipse:

$$A = \pi RL$$

$$I_C = \frac{AR^2}{4} = \frac{\pi R^3 L}{4}$$

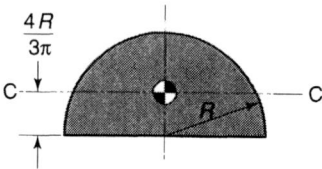

Semi-circle:

$$A = \frac{\pi R^2}{2}$$

$$I_C = \left(\frac{1}{4} - \frac{16}{9\pi^2}\right) AR^2$$

$$= \left(\frac{\pi}{4} - \frac{16}{9\pi}\right) \frac{R^4}{2}$$

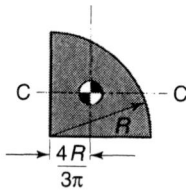

Quadrant of circle

$$A = \frac{\pi R^2}{4}$$

$$I_C = \left(\frac{1}{4} - \frac{16}{9\pi^2}\right) AR^2$$

$$= \left(\frac{\pi}{4} - \frac{16}{9\pi}\right) \frac{R^4}{4}$$

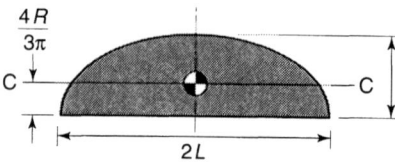

Semi-ellipse:

$$A = \frac{\pi RL}{2}$$

$$I_C = \left(\frac{1}{4} - \frac{16}{9\pi^2}\right) AR^2$$

$$= \left(\frac{\pi}{4} - \frac{16}{9\pi}\right) \frac{R^3 L}{2}$$

(Continued)

Table A.9 *(continued)*

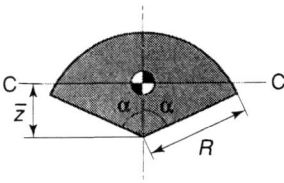

Circular sector:

$$\bar{z} = \frac{2R\sin\alpha}{3\alpha}$$

$$A = \alpha R^2$$

$$I_C = \left(\frac{\alpha + \sin\alpha\cos\alpha}{4} - \frac{4\sin^2\alpha}{9\alpha}\right)R^4$$

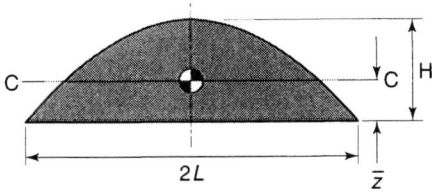

Sine wave:

$$\bar{z} = \frac{\pi H}{8}$$

$$A = \frac{4LH}{\pi}$$

$$I_C = \left(\frac{2}{9} - \frac{\pi^2}{64}\right)A^2$$

$$= \left(\frac{8}{9\pi} - \frac{\pi}{16}\right)LH^3$$

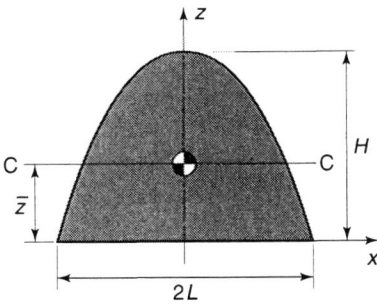

Segment of nth degree

$$z = H\left[1 - \left(\frac{x}{L}\right)^n\right]$$

$$\bar{z} = \frac{nH}{2n+1}$$

$$A = \frac{nLH}{n+1}$$

$$I_C = \frac{n^2(n+1)\,H^2A}{(2n+1)^2\,(3n+1)}$$

$$= \frac{n^3LH^3}{(2n+1)^2\,(3n+1)}$$

Bibliography

Any reader seeking other books on the subject of fluid mechanics can choose from hundreds, many on the shelves of university libraries and a significant number available in bookshops. The two in this brief bibliography are included because, of the many on my own bookshelves, they are the two I refer to over and over again. White's is an excellent book, lucidly written and covering a much wider range of topics than the present book. Fay's book also covers a wide range of topics but illustrates the power and elegance of a more mathematical approach. Ipsen's book on dimensional analysis is included because it includes the first description of which I am aware of the step-by-step method.

The remaining titles are essentially reference books. Cardarelli provides details of every system of units known to man. Kaye and Laby is a classic, first published in 1911 but revised many times since. It includes a wealth of data on the physical properties of fluids and other substances, and a great deal more besides. Emiliani's *Scientific Companion* is much more general but provides fascinating insights into the nature of the atmosphere, the hydrosphere and other aspects of our environment. Mayhew and Rogers is primarily a set of steam tables, but also includes information about other fluids and conversion factors between the SI and imperial systems of units as well as the properties of the standard atmosphere. Also included here are two British Standards, one concerned with the design and installation of orifice-plate, nozzle and Venturi-tube flowmeters to which we refer in Chapter 8, the other giving details of the SI system of units and its application.

BSI (1997) BS EN ISO 5167–1: 1997: *Measurement of fluid flow by means of differential devices. Part 1. Orifice plates, nozzles and Venturi tubes inserted in circular cross-section conduits running full.* BSI, London.

BSI (1993) BS 5555: 1993: *SI units and recommendations for the use of their multiples and of certain other units.* BSI London.

Cardarelli, Francois (1997) *Scientific Unit Conversion.* Springer-Verlag, London.

Emiliani, Cesare (1995) *The Scientific Companion* (2nd edition). John Wiley & Sons, New York.

Fay, James A. (1994) *Introduction to Fluid Mechanics.* MIT Press, Cambridge, Massachusetts.

Ipsen, D.C. (1960) *Units, Dimensions and Dimensionless Numbers.* McGraw-Hill, New York.

Kaye, G.W.C. and Laby, T.H. (1973) *Tables of Physical and Chemical Constants* (14th edition). Longman, London.

Mayhew, G.F.C. and Rogers, Y.R. (1980) *Thermodynamic and Transport Properties of Fluids. SI Units.* Blackwell, Oxford.

White, Frank M. (1994) *Fluid Mechanics* (3rd edition). McGraw-Hill, New York.

Index

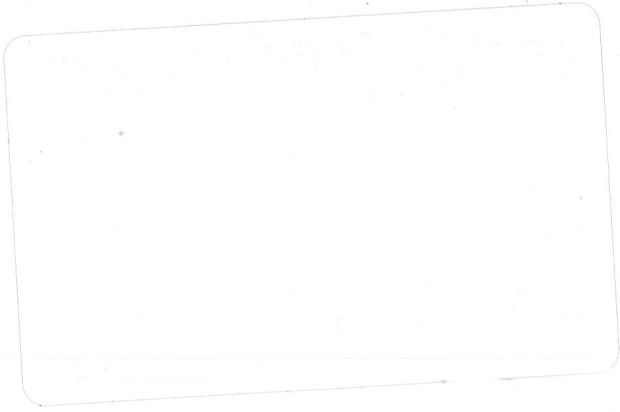